anatomy &

physiology

therapy basics

third edition

anatomy & physiology

therapy basics

third edition

Helen McGuinness

Hodder Arnold

A MEMBER OF THE HODDER HEADLINE GROUP

Orders: please contact Bookpoint Ltd, 130 Milton Park, Abingdon, Oxon OX14 4SB. Telephone: (44) 01235 827720. Fax: (44) 01235 400454. Lines are open from 9.00 to 5.00, Monday to Saturday, with a 24-hour message answering service. You can also order through our website www.hoddereducation.co.uk

If you have any comments to make about this, or any of our other titles, please send them to educationenquiries@hodder.co.uk

British Library Cataloguing in Publication Data
A catalogue record for this title is available from the British Library

ISBN-10: 0 340 908 084
ISBN-13: 978 0 340 908 082

First Edition Published 1995
Second Edition Published 2002
This Edition Published 2006
Impression number   7 6 5 4 3 2 1
Year   2006 2005 2004

Hodder Headline's policy is to use papers that are natural, renewable and recyclable products and made from wood grown in sustainable forests. The logging and manufacturing processes are expected to conform to the environmental regulations of the country of origin.

Cover photo from Ralph Mercer/The Image Bank/Getty Images.

Artwork by Cactus Design and Illustration Ltd

Picture credits: Science Photo Library – pp38, 39 (left), 51–58; Wellcome Photo Library – p39 (right), 47, 51 (bottom right)

Typeset in 10/12 Arrus BT by Charon Tec Ltd., Chennai, India, www.charontec.com

Printed in Great Britain for Hodder Arnold, an imprint of Hodder Education, a member of the Hodder Headline Group, 338 Euston Road, London NW1 3BH by CPI Bath

contents

acknowledgements

When I was preparing the original text of this book back in the early 1990s I never dreamed I would go on to write a third edition – and in full colour! I would therefore like to extend my grateful thanks to the following people who have helped me complete the considerable task of writing this full colour edition and accompanying CD-ROM.

To my husband Mark for his love, support and understanding, and especially for his excellent organisation and formatting of the text.

To my mum and dad for their love and constant words of encouragement and belief in my abilities.

To my dear friend Dee Chase for her constant love, support and encouragement throughout the revision of this book.

To Dr Nathan Moss for his help in checking the accuracy of the text.

I will always be greatly indebted to Deirdre Moynihan for her professional help and contributions throughout the preparation of the original text back in 1995.

To all the students and staff at the Holistic Training Centre, and all the colleges and lecturers who have used this book over the past ten years and who have been most encouraging and supportive of my work.

This book is devoted to my beautiful daughter Grace.

anatomical terminology

When studying anatomy and physiology, it is necessary to have a key or directional terminology to give precise descriptions when referring to body parts and structures. In anatomical terminology all parts of the body are described in relation to other body parts using a standardised body position called the **anatomical position**.

An anatomical position is determined from a central imaginary line running down the centre or midline of the body. In this position the body is erect and facing forwards, arms to the sides, palms facing forwards with the thumbs to the sides, and feet slightly apart with toes pointing forwards.

Learning anatomical terminology is like learning a new language. The terms you will need to become familiar with are as follows:

Anterior: front surface of the body, or structure

Posterior: back surface of the body, or structure

Deep: further from the surface

Superficial: near the surface

Internal: nearer the inside

External: nearer the outside

Lateral: away from the midline

Medial: towards the midline

Superior: situated above or towards the upper part

Inferior: situated below or towards the lower part

Proximal: nearest to the point of reference

Distal: furthest away from the point of reference

Prone: lying face down in a horizontal position

Supine: lying face up in a horizontal position

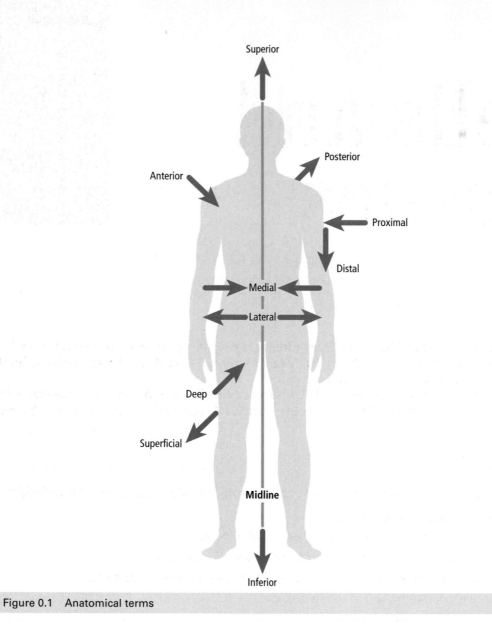

Figure 0.1   Anatomical terms

cells and tissues

Introduction

The human body can be likened to a universe, as it is made up of very small structures which are organised to function as a whole. The smaller structures are divided into five basic levels: chemical, cellular, tissues, organs and systems.

Ultimately all the body systems, and the minute cells that are the basic components of all organs and tissues, are involved in maintaining health and keeping the body in a state of balance.

Objectives

By the end of this chapter you will be able to recall and understand the following knowledge:

- the levels of organisation in the body
- parts of a cell's structure and their functional significance
- the structure and function of the main tissue types in the body
- the interrelationships between the cells and tissues and other body systems
- disorders associated with cells and tissues.

The levels of organisation in the body

The structure of the human body involves five principal levels of structural organisation:

■ **Atoms and molecules** – these represent the lowest level of organisational complexity in the body and are essential for maintaining life. At the chemical level the smallest unit of matter is the atom. An atom is the smallest particle of an element, such as a hydrogen or oxygen atom. A molecule is a particle composed of two or more atoms joined together, such as a water molecule (H_2O) or a carbon dioxide molecule (CO_2). Molecules combine to form cells.

■ **Cells** – these are the basic unit of all living organisms and are the smallest units that show characteristics of life.

■ **Tissues** – a tissue is a group of similar cells that perform a certain function (for example nervous and muscular tissues).

■ **Organs** – tissues are grouped into structurally and functionally integrated units called organs (for example the lungs or the heart).

■ **Systems** – a system is a group of organs that work together to perform specific functions. The systems of the body include the circulatory, skeletal, skin, respiratory, reproductive, muscular, endocrine, nervous, urinary and digestive systems.

Traditionally, the body is divided into different systems according to their specific functions. However, the ultimate purpose of each system is to maintain a constant internal environment for each cell to enable it to survive.

The human body is exposed to a constantly changing external environment. These changes are neutralised by the internal environment of blood, lymph and tissue fluids that bathe and protect the cells.

Body parts function efficiently only when the concentrations of water, food substances, oxygen and the conditions of heat and pressure remain within certain narrow limits. The process by which the body maintains a stable internal environment for its cells and tissues is called **homeostasis**.

The body and its systems are constructed in such a way that all systems work synergistically with one another with one overall aim – to maintain homeostasis. Examples of homeostatic mechanisms in the body include the regulation of body temperature, blood pressure and blood sugar levels. Homeostasis is maintained by adjusting the metabolism of the body.

Metabolism is the term used to describe the physiological processes that take place in our bodies to convert the food we eat and the air we breathe into the energy we need to function. Metabolism is essentially the basic chemical working of the body cells, and through metabolism food substances are transformed into energy or materials that the body can use or store.

Metabolism involves two processes:

■ **Catabolism** – the chemical breakdown of complex substances by the body to form simpler ones, accompanied by the release of energy. The substances broken down include nutrients in food (carbohydrates and proteins) as well as the body's storage products (glycogen).

■ **Anabolism** – the building up of complex molecules, such as proteins and fats, from simpler ones by living things. The rate at which a person consumes energy in activities and body processes is known as the **metabolic rate**. The minimum energy required to keep the body alive is known as the **basal metabolic rate**.

Cells

The cell is the fundamental unit of all living organisms and is the simplest form of life that can exist as a self-sustaining unit. Cells are therefore the building blocks of the human body.

Cells in the body take many forms, the size and shape being largely dependent on their specialised function. For example, some cells help fight disease; others transport oxygen or produce movement; some manufacture proteins and chemicals; and others store nutrients.

Cells consist of four elements: carbon, oxygen, hydrogen and nitrogen, as well as trace elements such as iron, sodium and potassium. Trace elements are of significance in certain cellular functions, for example calcium is needed for blood clotting. Besides the four primary elements, water makes up 60 to 80 per cent of all cells.

Cell organelles

Molecules combine in very specific ways to form what is called cell organelles (little organs) which are the basic component parts of the cells. Each organelle has a particular functional significance within the cell that allows it to live.

Typical cell organelles include the **cell membrane, cytoplasm, nucleus, nucleolus, nuclear membrane, lysosome, vacuole, ribosome, Golgi body, mitochondria, centrosome, centromere, centrioles, chromatin, endoplasmic reticulum** and **chromatids**. Despite the great variety of cells in the body, they all have the same basic structure.

Cell membrane

The cell membrane is a fine membrane that encloses the cell and protects its contents. This membrane is said to be semipermeable in that it selectively controls the inward and outward movement of molecules into and out of the cell. Oxygen, nutrients, hormones and proteins are taken into the cell, as needed, and cellular waste such as carbon dioxide passes out through the membrane. As well as governing the exchange of nutrients and waste materials, its function is also to maintain the shape of the cell.

Cytoplasm

The cytoplasm is the gel-like substance that is enclosed by the cell membrane. The cytoplasm contains the nucleus and the small cellular structures called organelles. Most cellular metabolism occurs within the cytoplasm of the cell.

Nucleus

The nucleus is the largest organelle in the cytoplasm and is the control centre of the cell, regulating the cell's functions and directing most metabolic activities. The nucleus governs the specialised work performed by the cell and the cell's own growth, repair and reproduction. All cells have at least one nucleus at some time in their existence. The nucleus is significant in that it contains all the information required for the cell to function and controls all cellular operations.

The information required by the cell is stored in DNA (deoxyribonucleic acid) which carries the genetic materials for replication of identical molecules. The DNA strands are found in threadlike structures known as chromosomes. Each human cell has 23 pairs of chromosomes.

Chromatin

Chromatin is the substance inside the nucleus that contains the genetic material.

Nucleolus

Inside the nucleus is a dense, spherical structure called a nucleolus which contains ribonucleic acid (RNA) structures that form ribosomes.

Nuclear membrane

The nucleus is surrounded by a perforated outer membrane called the nuclear membrane; materials move across it to and from the cytoplasm.

Key note

DNA is a long, twisted molecule found in the chromatin of the cell's nucleus. It is often called the body's blueprint, as it is a record of a person's height, bone structure, hair colour, body chemistry and other characteristics. When cells divide and multiply, DNA makes sure that the new cells are direct copies by passing on its hereditary information.

Centrosome

This is an area of clear cytoplasm found next to the nucleus, and contains the centrioles.

Centrioles

Contained within the centrosomes are the small spherical structures called centrioles which are associated with cell division, or mitosis. During cell division the centriole divides in two and migrates to opposite sides of the nucleus to form the spindle poles.

Chromatids

This is a pair of identical strands that are joined at the centromere and separate during cell division.

Centromere

The portion of a chromosome where the two chromatids are joined is the centromere.

Key note

A **chromosome** is one of the thread-like structures in the cell nucleus that carry the genetic information in the form of genes. The nucleus of a human cell contains 46 chromosomes, 23 of which are maternal and 23 of which are of paternal origin. Each chromosome can duplicate an exact copy of itself between each set cell division so that each new cell formed receives a full set of chromosomes.

Ribosomes

Ribosomes are tiny organelles made up of ribonucleic acid (RNA) and protein. They may be fixed to the walls of the endoplasmic reticulum (known as rough ER) or may float freely in the cytoplasm. Their function is to manufacture proteins for use within the cell and also to produce other proteins that are exported outside the cell.

Endoplasmic reticulum

This is a series of membranes continuous with the cell membrane. It can be thought of as an intracellular transport system, allowing movement of materials from one part of the cell to another. It links the cell membrane with the nuclear membrane and therefore assists movement and materials out of the cell.

It contains enzymes and participates in the synthesis of proteins, carbohydrates and lipids. The endoplasmic reticulum serves to store material, transport substances inside the cell, as well as detoxify harmful agents. Some endoplasmic reticulum appears smooth, but it can also appear rough due to the presence of ribosomes.

Mitochondria

Mitochondria are oval-shaped organelles that lie in varying numbers within the cytoplasm. The mitochondria are said to be the site of the cell's energy production. The mitochondria provides most of a cell's ATP (adenosine triphosphate) which is a compound that stores the energy needed by the cell.

The work of the mitochondria is assisted by enzymes which are proteins that speed up chemical changes. By means of cellular respiration, the mitochondria provide the energy which powers the cell's activities.

Lysosome

These are round sacs present in the cytoplasm. They contain powerful enzymes, which are capable of digesting proteins. Their function is to destroy any part of the cell that is worn out so that it can be eliminated – this is known as lysis.

Vacuole

These are empty spaces within the cytoplasm. They contain waste materials or secretions formed by the cytoplasm and are used for temporary storage, transportation or digestive purposes in different kinds of cells.

Golgi body/apparatus

This is a collection of flattened sacs within the cytoplasm. The Golgi apparatus is typically located near the nucleus and attached to the endoplasmic reticulum. It is said to be the 'packaging department' of the cell as it stores the protein manufactured in the endoplasmic reticulum and later transports it out of the cell.

Functions of cells

In order for a cell to survive it must be able to carry out a variety of functions.

Respiration

Every cell requires oxygen for the process of metabolism. Oxygen is absorbed through the cell's semipermeable membrane and is used to oxidise nutrient material to provide heat and energy. The waste products produced as a result of cell respiration include carbon dioxide and water. These are passed out from the cell through its semipermeable membrane.

Growth

Cells have the ability to grow until they are mature and ready to produce. A cell can grow and repair itself by manufacturing protein.

Excretion

During metabolism various substances are produced which are of no further use to the cell. These waste products are removed through the cell's semipermeable membrane.

Movement

Movement may occur in the whole or part of a cell. White blood cells, for instance, are able to move freely.

Irritability

A cell has the ability to respond to a stimulus, which may be physical, chemical or thermal. For example, a muscle fibre contracts when stimulated by a nerve cell.

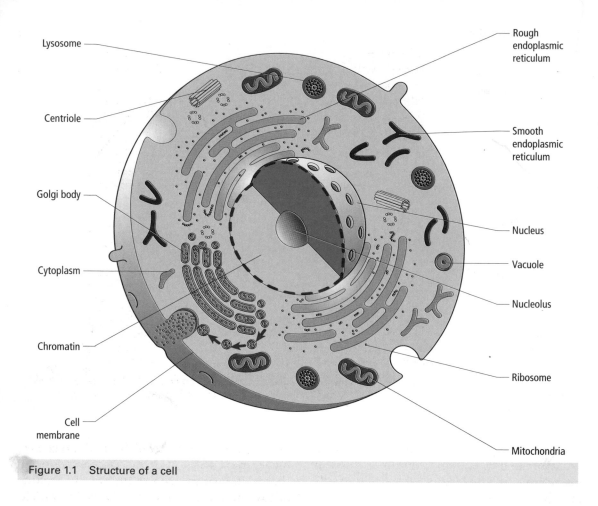

Figure 1.1 Structure of a cell

Reproduction

When growth is complete in a cell, reproduction takes place. The cells of the human body reproduce or divide by the process of mitosis.

Cellular respiration

In order to function properly a cell must maintain a stable internal environment and therefore the transport of materials has to be achieved without an excessive build-up of chemicals. The term cell respiration refers to the controlled exchange of nutrients (such as oxygen and glucose) and waste (such as carbon dioxide) by the cell to activate the energy needed for the cell to function.

In order for cells to carry out their work they need to produce enough energy or fuel. Fuel is provided by glucose from carbohydrate metabolism and in order for the glucose to be released or 'oxidised', oxygen is absorbed from the respiratory system into the bloodstream.

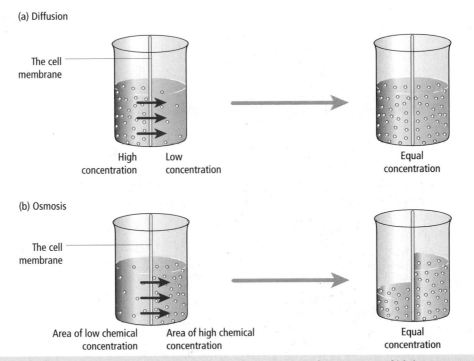

(a) Diffusion

The cell membrane

High concentration   Low concentration

Equal concentration

(b) Osmosis

The cell membrane

Area of low chemical concentration   Area of high chemical concentration

Equal concentration

Figure 1.2   (a) Diffusion: the process in which small molecules move from an area of high concentration to lower concentration; (b) Osmosis: the movement of water through the cell membrane from low to high chemical concentration

Cells are bathed in a fluid called tissue fluid, or interstitial fluid, which allows the interchange of substances between the cells and the blood, known as internal respiration.

The body's internal transport system, the blood, carries oxygen from the respiratory system and nutrients such as glucose from the digestive system to the cells, and these are absorbed through the cell membrane in several different ways: **diffusion, osmosis, active transport** and **filtration**. When certain molecules are needed, such as glucose, the cell will take these in and discard other materials in order to preserve the equilibrium.

Diffusion

As chemicals become concentrated outside the cell a flow of small molecules takes place through the cell membrane until a balance exists. This process in which small molecules move from areas of high concentration to those of lower concentration is called diffusion. Diffusion is the basis by which the cells lining the small intestines take in digestive products to be utilised by the body.

Osmosis

This process refers to the movement of water through the cell membrane from areas of low chemical concentration to areas of high chemical concentration. This process allows for the dilution of chemicals, which are unable to cross the cell membrane by diffusion, in order to maintain equilibrium within the cell.

Active transport

This is an energy-dependent process in which certain substances (including ions, some drugs and amino acids) are able to cross cell membranes against a concentration gradient. This is the process, using chemical energy, by which the cell takes in larger molecules that would otherwise be unable to enter in sufficient quantities. Carrier molecules within the cell membrane bind themselves to the incoming molecules, rotate around them and release them into the cell. This is the means by which the cell absorbs glucose.

Filtration

This is the movement of water and dissolved substances across the cell membrane due to differences in pressure. The force of the weight of the fluid pushes against the cell membrane, thereby moving it into the cell. One site of filtration in the body is in the kidneys. Blood pressure forces water and small molecules through plasma membranes of cells and the filtered liquid then enters the kidneys for filtration.

The cell's life cycle

It is vital for living cells to reproduce themselves in order to continue life and cells undergo many divisions from the time of fertilisation to physical maturity. When a single cell undergoes division, it forms two daughter cells that are identical to the original cell. A cell may live from a few days to many years, depending on its type. Cells divide in two ways: **mitosis** and **meiosis**.

Mitosis

Mitosis is when a single cell produces two genetically identical daughter cells. It is the way in which new body cells are produced for both growth and repair. Division of the nucleus takes place in four phases (prophase, metaphase, anaphase and telophase) and is followed by the division of the cytoplasm to form the daughter cells.

Prophase

- Centrioles duplicate as chromatids in the nucleus change to become individual chromosomes.

- Centrioles separate and form spindles.

Metaphase

- Chromosomes align themselves in the centre of the cell, midway between the centrioles as the nucleus and its protective membrane disappears.

- The centromere of each chromosome then replicates.

Anaphase

- Centromeres divide and identical sets of chromosomes move to opposite poles of the cell.

Telophase

- This is the final stage of mitosis.

- Nuclear membrane forms around each nucleus and spindle fibres disappear.

- Cytoplasm compresses and divides in half.

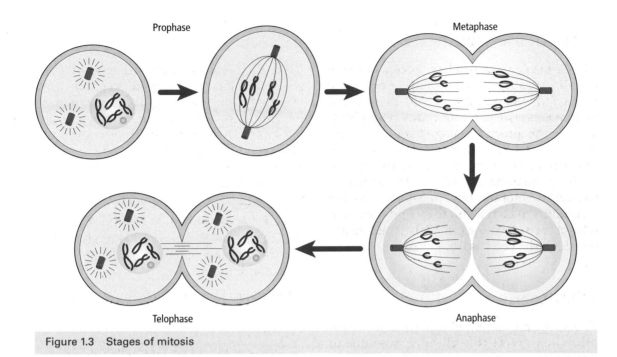

Prophase

Metaphase

Telophase

Anaphase

Figure 1.3   Stages of mitosis

Meiosis

Meiosis is a type of cell division that produces four daughter cells, each having half the number of chromosomes of the original cell. Meiosis involves the production of a new organism, formed by the fusion of a sperm from the male and an egg from the female.

Before fertilisation there are only 23 chromosomes present in the sperm and the egg. After fertilisation has taken place the egg and the sperm fuse together to form a single cell called a zygote with 46 chromosomes (23 from each parent). The zygote is then able to reproduce itself by cell division or mitosis to form an embryo, foetus and eventually a fully formed individual.

Tissues

Tissues are defined as a group of similar cells that act together to perform a specific function. The study of tissues is known as **histology**.

Due to the complexity of the human body it is not possible for every cell to carry out all the functions required by the body. Some cells, therefore, become specialised to form a group of cells or tissues. It is important for a therapist to have a basic understanding the tissues in order to understand the structure and functions of the body's organs.

There are four major types of tissues in the human body:

- **epithelial** tissue
- **connective** tissue
- **muscle** tissue
- **nervous** tissue.

All four types of tissue have special purposes and therefore have varying different rates of cellular regeneration. Epithelial tissue is renewed constantly by the process of cell division or mitosis.

Bone tissue and adipose connective tissue are highly vascular and therefore heal quickly. Muscle tissue takes longer to regenerate. Nervous tissue regenerates very slowly. The less vascular forms of connective tissue, such as ligaments and tendons, are even slower to heal than muscle tissue, and cartilage is among the slowest to heal.

Epithelial tissue

Epithelial tissue consists of sheets of cells which cover and protect the external and internal surfaces of the body and line the inside of hollow structures. They specialise in moving substances in and out of the blood during secretion, absorption and excretion. As they are subject to a considerable amount of wear and tear, epithelial cells reproduce very actively.

Usually there is little matrix present in epithelial tissues. The matrix present tends to form continuous sheets of cells, with the cells held very close together. A thin, permeable basement membrane attaches epithelial tissues to the underlying connective tissue. Epithelial tissue, which consists of cells closely packed together, come in various shapes. There are two categories of epithelial tissue:

- **simple** (single-layered)

- **compound** (multilayered).

Simple epithelium

Simple epithelium tissues have only one layer of cells over a basement membrane. Being thin, they are fragile and are found only in areas inside the body which are relatively protected, such as the lining of the heart and blood vessels and the lining of body cavities. They are also found lining the digestive tract and in the exchange surfaces of the lungs where their thinness is an advantage for speedy absorption across them. There are four different types of simple epithelium, named according to their shape and the functions they perform.

- **Squamous epithelium** – these cells are flat, scalelike cells with a central nucleus. The cells fit closely together, rather like a pavement, producing a very smooth surface. As it is so flat, this type of tissue is particularly well suited to cellular processes such as diffusion. Squamous epithelium lines the alveoli of the lungs. It is also found lining blood vessels and the heart, where it is known as endothelium.

- **Cuboidal epithelium** – the cube-like shape of these cells is only visible when the tissue is sectioned through at right angles. Cuboidal epithelium is found in areas where absorption or secretion takes place, such as the ovaries, kidney tubules, thyroid gland, pancreas and salivary glands.

- **Columnar epithelium** – these cells are much higher than they are wide and consist of a single layer of cells of cylindrical shape, with the nucleus situated towards the base of the cell. This type of tissue lines the small and large intestine, stomach and gall bladder and is involved in secretion and absorption. Its thickness helps to protect underlying tissues and many of these types of cells are modified for a particular function. For example, in the small intestine the plasma membrane is folded into microvilli whose function is the absorption of nutrients.

- **Ciliated epithelium** – this is a form of columnar epithelium and has hairlike projections from its surface. This type of cell lines the respiratory system. The beating of the cilia carries unwanted particles along with mucus out of the system and maintains cleanliness. It also lines the uterine tubes to help propel the ova towards the uterus.

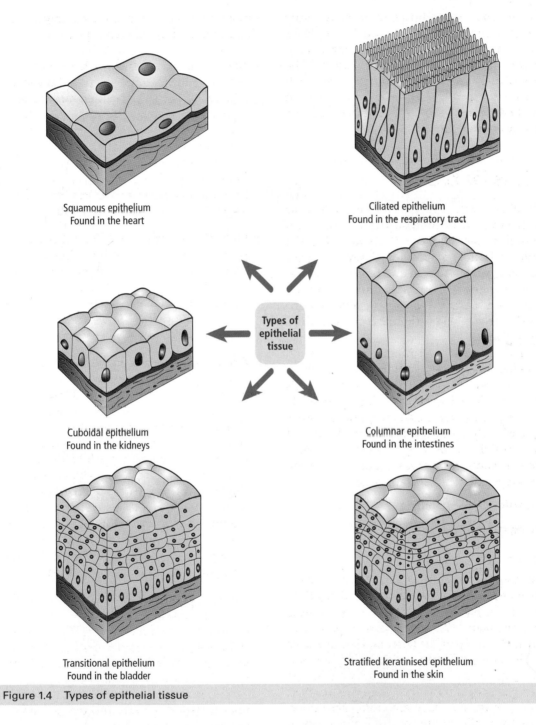

Squamous epithelium
Found in the heart

Ciliated epithelium
Found in the respiratory tract

Types of epithelial tissue

Cuboidal epithelium
Found in the kidneys

Columnar epithelium
Found in the intestines

Transitional epithelium
Found in the bladder

Stratified keratinised epithelium
Found in the skin

Figure 1.4   Types of epithelial tissue

Compound epithelium

The main function of compound epithelium is to protect underlying structures. Compound epithelium contains two or more layers of cells. There are two main types:

- **stratified**   - **transitional.**

Stratified epithelium is composed of a number of layers of cells of different shapes. In the deeper layers the cells are mainly columnar in shape and as they grow towards the surface they become flattened. There are two types of stratified epithelium:

- **Non-keratinised stratified epithelium** – this is found on wet surfaces that may be subject to wear and tear, such as the conjunctiva of the eyes, the lining of the mouth, the pharynx and the oesophagus.

- **Keratinised stratified epithelium** – this is found on dry surfaces such as the lining the skin, hair and nails. The surface layers of keratinised cells are dead cells. They give protection and prevent drying out of the cells in the deeper layers from which they develop. The surface layer of cells is continually being rubbed off and is replaced from below.

Transitional epithelium is composed of several layers of pear-shaped cells which change shape when they are stretched. This type of tissue is found lining the uterus, bladder and pelvis of the kidney.

Connective tissue

Connective tissue is the most abundant type of tissue in the body. It connects tissues and organs by binding the various parts of the body together and helps to give protection and support. Connective tissues are made up of cells and matrix, which is like the ground substance.

Connective tissue cells are often more widely separated from each other than those forming epithelial tissue, and the space between cells is filled with a large amount of non-living matrix. There may or may not be fibres in the matrix, which may be either of a semi-solid, jellylike consistency or dense and rigid, depending on the position and function of the tissue.

There are several different types of connective tissue.

Areolar tissue

This is the most widely distributed type of connective tissue in the body. This type of tissue is composed of cells called fibrocytes which are widely separated by white and reticular fibres, as well as yellow elastic fibres.

This tissue allows for elasticity and is found in almost every part of the body connecting and supporting organs: under the skin, between muscles, supporting blood vessels and nerves and in the alimentary canal.

Adipose tissue

This tissue is composed of specialised cells for the storage of fat – adipocytes. These cells are present within the matrix but few fibres are present. Its function is to provide protection to the organs close to it, to reduce heat loss and to act as an emergency energy reserve.

This type of tissue is found supporting organs such as the kidneys and eyes, between bundles of muscle fibres, in the yellow bone marrow of long bones and as a padding around joints. Along with areolar tissue, adipose tissue is found under the skin, in the subcutaneous layer, giving the body a smooth, continuous line.

White fibrous tissue

This is a strong, connecting tissue made up of mainly closely packed bundles of white, collagenous fibres, with very little matrix.

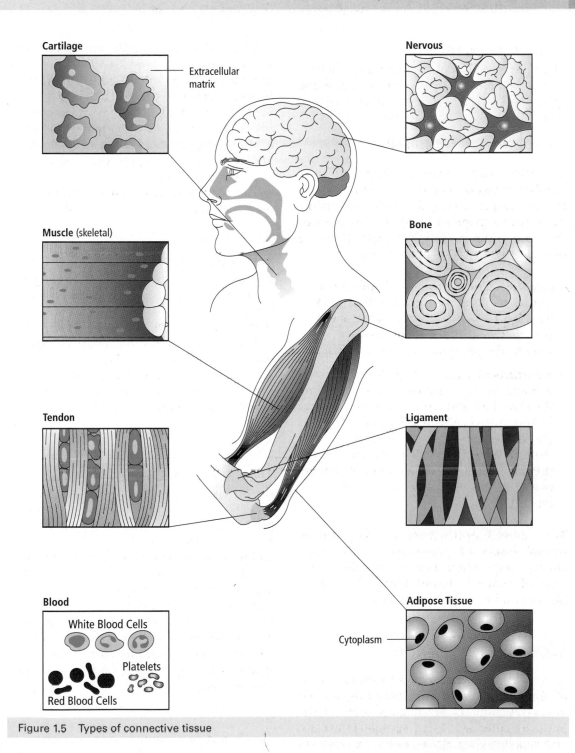

Figure 1.5 Types of connective tissue

Cells called fibrocytes are present between the bundles. This type of connective tissue forms tendons which attach muscle to bone, ligaments which tie bones together and as an outer protective covering for some organs such as the kidney and bladder. The function of this tissue is to provide strong attachment between different structures.

Yellow elastic tissue

This consists of branching yellow elastic fibres, with fibrocytes in the spaces between the fibres. This tissue is found in areas where alteration of shape is needed, such as the arteries, trachea and bronchi and lungs. Its function is to allow the stretching of various organs, followed by a return to their original shape and size.

Lymphoid tissue

This tissue has a semi-solid matrix with fine branching fibres. The cells contained within this tissue are specialised and are called lymphocytes. It is found in the lymph nodes, spleen, tonsils, adenoids, walls of the large intestine and glands in the small intestine. This type of tissue forms part of the lymphatic system whose function is to protect the body from infection.

Blood (fluid or liquid tissue)

Blood is also known as liquid connective tissue. It contains the blood cells erythrocytes, leucocytes and thrombocytes, which float within a fluid called plasma. Blood helps maintain homeostasis of the body by transporting substances throughout the body, by resisting infection and by maintaining heat.

Bone tissue

Bone is the hardest and most solid of all connective tissues. It consists of compact tissue, which is a dense type designed for strength; spongy cancellous tissue for structure bearing; collagenous fibres for strength; and mineral salts for hardness. Bone is infused with many blood vessels and nerves through a membranous sheath surrounding the bone called the periosteum.

Cartilage

This is a much firmer tissue than any of the other connective tissues and the matrix is quite solid. For descriptive purposes, cartilage is divided into three types:

- **hyaline** cartilage
- **white fibrous** cartilage
- **yellow elastic** fibrocartilage.

Hyaline cartilage is a smooth, bluish-white, glossy tissue. It contains numerous cells called chondrocytes from which cartilage is produced. Hyaline cartilage is the most abundant type and is found on the surfaces of the parts of bones which form joints, forming the costal cartilage which attaches the ribs to the sternum, forming part of the larynx, trachea and bronchi.

This type of cartilage provides a hard-wearing, low-friction surface within joints and is flexible to provide support in, for example, the nose and trachea.

White fibrous cartilage is composed of bundles of collagenous white fibres in a solid matrix with cells scattered among them. This type of cartilage is tough but slightly flexible. It is found as pads between the bodies of the vertebrae called the intervertebral discs and in the symphysis pubis which joins the pubis bones together. Its function is one of support and to join together or fuse certain bones.

Yellow elastic fibrocartilage consists of yellow elastic fibres running through a solid matrix, between which chondrocytes are situated. It is found forming the pinna (lobe of the ear) and the epiglottis. The function of this type of cartilage is to provide support and to maintain shape.

Muscle tissue

Muscle tissue is very elastic and therefore has the unique ability to provide movement by shortening as a result of contraction. This tissue is made up of contractile fibres, usually arranged in bundles and surrounded by connective tissue. There are three types of muscle tissue:

- **voluntary** (skeletal) tissue
- **Involuntary** or smooth tissue
- **cardiac muscle** tissue.

The different types of muscle tissue are discussed in more detail in Chapter 4 – the muscular system.

Nervous tissue

Nervous tissue consists of cells called neurones which can pick up and transmit electrical signals by converting stimuli into nerve impulses. Nervous tissue has the characteristics of excitability and conductivity. Its functions are to coordinate and regulate body activity.

Nervous tissue and neurones are discussed in more detail in Chapter 8 – the nervous system.

Membranes

Membranes are a thin, soft, sheetlike layer of tissue that covers a cell, organ or structure, that lines tubes or cavities or divides and separates one part of a cavity from another. There are three basic types of membranes in the body:

- **mucous** membrane
- **serous** membrane
- **synovial** membrane.

Mucous membrane

Mucous membrane is found lining the opening to the external environment, such as those of the respiratory, digestive, urinary and reproductive tracts. Mucous membrane secretes a viscous, slippery fluid called mucus that coats and protects the underlying cells.

Serous membrane

Serous membranes line body cavities that are not open to the external environment and cover many of the organs. These membranes consist of two layers: a parietal layer which lines the wall of body cavities, and a visceral layer which provides an external covering to organs in body cavities. They secrete a thin, watery fluid that lubricates organs to reduce friction as they rub against one another and against the wall of the cavities. Examples of serous membranes include the pericardium of the heart, pleural membranes in the lungs and the peritoneum lining the abdominal organs.

Synovial membrane

Synovial membrane lines the joint cavities of freely movable joints such as the shoulder, hip and knee. These membranes secrete synovial fluid that provides nutrition and lubrication to the joint so that it can move freely without undue friction. This type of membrane is also found in bursae, which are protective sacs located around a joint cavity, between layers of muscle and connective tissue and wherever the body needs extra protection.

Disorders of cells and tissues

Cancer/abnormal cell division

Cancerous diseases are characterised by the growth of abnormal cells that invade surrounding tissues and **metastasis** (the spread of cancerous cells to other parts of the body).

When cells in an area of the body divide without control, the excess tissue that develops is called a **tumour**, growth or **neoplasm**. The study of tumours is called **oncology** and a physician who specialises in this field is called an **oncologist**.

Tumours may be cancerous and sometimes fatal, or they may be quite harmless. A cancerous growth is called a **malignant** tumour and a non-cancerous growth is called a **benign** tumour. Benign tumours do not spread to other parts of the body, but they may be removed if they interfere with a normal body function or are disfiguring. Cancers are classified according to their microscopic appearance and the body site from which they arise.

The name of the cancer is derived from the type of tissue in which it develops. Most human cancers are **carcinomas**, malignant tumours that arise from epithelial cells. **Melanomas** are cancerous growths of melanocytes, the skin cells that produce the pigment melanin.

Sarcoma is a general term for any cancer arising from muscle cells or connective tissues. For example, osteogenic sarcoma (bone cancer) is the most frequent type of childhood cancer which destroys normal bone tissue and eventually spreads to other areas of the body.

Leukaemia is a cancer of blood-forming organs characterised by rapid growth and distorted development of leucocytes. **Lymphoma** is a malignant disease of lymphatic tissue such as the lymph nodes. An example is Hodgkin's disease.

Interrelationships with other systems

Cells and tissues link to the following body systems.

Skin

Keratinised stratified epithelium (a type of tissue containing layers of cells) is found on dry surfaces such as the skin, hair and nails.

Skeletal

Bone is the hardest and most solid type of connective tissue in the body which is needed for building the structures of the skeletal framework.

Muscular

There are three types of muscle tissue: skeletal muscle which controls voluntary movements; smooth muscle which controls involuntary movements, and cardiac muscle which controls the heart.

Circulatory

Blood is a form of liquid connective tissue whose role is in transporting substances to and from the cells.

Respiratory

A type of tissue called ciliated columnar epithelium lines the respiratory tract, which carries unwanted particles out of the system.

Nervous

Neurones and neuroglia are the specialised cells that form nervous tissue which enables the body to receive and transmit nerve impulses in order to regulate and coordinate body activities.

Endocrine

The endocrine glands are made from epithelial tissue. They secrete hormones directly into the bloodstream to influence the activity of another organ or gland.

Digestive

The digestive system is lined with epithelial tissue with goblet cells that secrete mucus to aid the flow of the digestive processes.

Urinary

The bladder is lined with transitional epithelium which allows the bladder to expand when full and deflate when empty.

Key words associated with cells

atom	nucleus	centrioles
molecule	nucleolus	chromatid
cell	nuclear membrane	diffusion
tissue	lysosome	osmosis
organ	vacuole	active transport
system	ribosome	filtration
homeostasis	golgi body	cell respiration
metabolism	mitochondria	tissue fluid
cell membrane	centrosome	meiosis
cytoplasm	centromere	mitosis

Summary of cells

- The human body involves five levels of structural organisation – atoms and molecules, cells, tissues, organs and systems.

- **Atoms and molecules** are the lowest level of organisational complexity in the body.

- **Cells** are the smallest units that show characteristics of life.

- **Tissues** are a group of similar cells that perform a certain function.

- **Organs** are tissues grouped into structurally and functionally integrated units.

- **Systems** are a group of organs that work together to perform specific functions.

- The process by which the body maintains a stable internal environment for its cells and tissues is called **homeostasis**.

- **Metabolism** is the term used to describe the physiological processes that take place in our bodies to convert the food we eat and the air we breathe into the energy we need to function.

- The minimum energy required to keep the body alive is known as the **basal metabolic rate**.

- A **cell** is the basic, living, structural and functional unit of the body.

- The principal parts of the cell are the **cell membrane** and its **organelles** which play specific roles in cellular growth, maintenance, repair and control.

- The **cell membrane** encloses the cell and protects its contents. It is semipermeable and governs the exchange of nutrients and waste materials.

- The **nucleus** controls the cell's activities and contains the genetic information.

- The **cytoplasm** is the substance inside the cell between the plasma membrane and the nucleus.

- The **ribosomes** are sites of protein synthesis.

- The **endoplasmic reticulum** links the cell membrane with the nuclear membrane and assists movement of materials out of the cell.

- The **Golgi body** processes, sorts and delivers proteins and lipids (fats) to the plasma membrane, lysosome and secretory vesicles.

- The **lysosome** is a round sac in the cytoplasm that contains powerful enzymes to help destroy waste and worn-out cell materials.

- The **mitochondria** are the 'powerhouses' of the cell.

- The **centrosome** is a dense area of cytoplasm containing the centrioles.

- The **centrioles** are paired, small, spherical structures associated with cell division, or mitosis.

- The **chromatids** are a pair of identical strands that are joined at the centromere and separate during cell division.

- The **centromere** is the portion of a chromosome where the two chromatids are joined.

- Cells function through the exchange of fluids, nutrients, chemicals and ions which are carried out by passive processes such as **diffusion**, **osmosis** and **filtration**, and active processes such as **active transport**.

- **Cell respiration** is the controlled exchange of nutrients, such as oxygen and glucose, and waste, such as carbon dioxide, by the cell to activate the energy needed for the cell to function.

- The fuel required by cells is provided by **glucose** from carbohydrate metabolism and **oxygen** absorbed from the respiratory system into the bloodstream.

- Cells are bathed in a fluid known as **tissue fluid** or interstitial fluid which allows the interchange of substances between the cells and the blood, known as internal respiration.

- Cell division is the process by which cells reproduce themselves.

- **Mitosis** is cell division that results in an increase in body cells and involves division of a nucleus.

- **Meiosis** is reproductive cell division and results in the fusion of an egg and a sperm into a zygote.

Key words associated with tissues

tissue	cartilage
epithelial tissue	muscular tissue
simple epithelium	nervous tissue
compound epithelium	squamous epithelium
connective tissue	cuboidal epithelium
areolar tissue	columnar epithelium
adipose tissue	ciliated epithelium
white fibrous tissue	stratified epithelium
yellow elastic tissue	transitional epithelium
lymphoid tissue	mucous membrane
blood	serous membrane
bone	synovial membrane

Summary of tissues

- A **tissue** is a group of similar cells that are specialised for a particular function.

- The tissues of the body are classified into four main types: **epithelial, connective, muscular** and **nervous**.

- **Epithelial tissue** provides coverings and linings of many organs and vessels.

- There are two categories of epithelial tissue: **simple** (single-layer) and **compound** (multilayer).

- There are four different types of simple epithelium: **squamous, cuboidal, columnar** and **ciliated**.

- There are two different types of compound epithelium: **stratified and transitional**.

- **Connective tissue** is the most abundant type of body tissue. It connects tissues and organs to give protection and support.

- Connective tissue consists of the following different types: **areolar,**

- **adipose, white fibrous, yellow elastic, lymphoid, blood, bone** and **cartilage**.

- **Muscle tissue** is elastic and is therefore modified for contraction. It is found attached to bones (skeletal muscle), in the wall of the heart (cardiac muscle) and in the walls of the stomach, intestines, bladder, uterus and blood vessels.

- **Nervous tissue** is composed of nerve cells called neurones which pick up and transmit nerve signals. Membranes are thin, soft, sheetlike layers of tissue.

- **Mucous membrane** lines cavities that open to the exterior such as the digestive tract.

- **Serous membrane** lines body cavities that are not open to the external environment (the lungs and the heart).

- **Synovial membrane** lines joint cavities of freely movable joints such as the shoulder, hip and knee.

Multiple-choice questions

1. **The study of the structures of the human body and their relationship to one another is known as:**

   a homeostasis

   b metabolism

   c anatomy

   d physiology

2. **The process by which the body maintains a stable internal environment of its cells and tissues is:**

   a physiology

   b metabolism

   c homeostasis

   d anatomy

3. **The simplest form of life that can exist as independent self-sustaining units are:**

   a tissues

   b cells

   c atoms

   d organs

4. **The correct order of the five levels of organisation in the human body from the lowest to the highest is:**

   a cells, atoms and molecules, organs, tissues, systems

   b tissues, cells, atoms and molecules, organs, systems

   c atoms and molecules, cells, tissues, organs, systems

   d atoms and molecules, tissues, cells, organs, systems

5. **The process by which new body cells are produced for both growth and repair is called:**

   a meiosis

   b metaphase

   c mitosis

   d none of the above

6. **The control centre of the cell that directs nearly all metabolic activities is the:**

   a mitochondria

   b Golgi body

   c nucleus

   d cell membrane

7. **The organelle that powers the cell activities is the:**

   a nucleus

   b lysosome

   c mitochondria

   d endoplasmic reticulum

8. **The small spherical structures associated with cell division are:**

   a centrosomes

   b centrioles

   c chromatids

   d centromeres

9. **The process in which small molecules move from areas of high concentration to those of lower concentration is:**

   a osmosis

   b diffusion

   c filtration

   d active transport

10. **A group of similar cells that group together to perform a certain function are known as:**

a   tissues
b   organs
c   cells
d   atoms

11. **The type of tissue that lines the internal and external organs of the body and lines vessels and body cavities is:**

a   connective tissue
b   epithelial tissue
c   serous tissue
d   nervous tissue

12. **Where would you find ciliated epithelium?**

a   in the kidney tubules
b   in the eyes
c   in the respiratory system
d   in the pancreas

13. **Where would you find columnar epithelium?**

a   in the small and large intestine
b   in the stomach
c   in the gall bladder
d   all of the above

14. **The most widely distributed type of connective tissue in the body is:**

a   areolar tissue
b   adipose tissue
c   epithelial tissue
d   white fibrous tissue

15. **The most abundant type of cartilage found on the surface of parts of bones which form joints is:**

a   elastic cartilage
b   fibrocartilage
c   yellow elastic fibrocartilage
d   hyaline cartilage

16. **Membranes that line openings to the outside of the body are called:**

a   serous membranes
b   mucous membranes
c   synovial membranes
d   epithelial membranes

the skin

Introduction

The skin is one the largest organs in the body in terms of weight and surface area. It can be classified as an appendage as it consists of different types of tissues that are joined to perform specific functions.

The nail is an appendage of the skin and is a modification of the stratum corneum (horny) and stratum lucidum (clear) layers of the epidermis. Nails are non-living tissue. Their two main functions are protection for the fingers and toes and as tools for the manipulation of objects. The hair is also an appendage of the skin and grows from a sac-like depression in the epidermis called a hair follicle. The primary function of hair is protection.

Objectives

By the end of this chapter you will be able to recall and understand the following knowledge:

- the structure and functions of the skin and its appendages
- the characteristics of ethnic skin types
- the interrelationships between the skin and other body systems
- knowledge of skin, hair and nail diseases and disorders.

The skin

The skin is like a cell membrane, defining our parameters. Located within its layers are several types of tissues that carry out special functions such as protection, temperature regulation and excretion. The skin also forms natural openings for the mouth, nose and parts of the urino-genital systems. No other body system is more easily exposed to infections, disease, pollution or injury than the skin, yet no other body system is as strong and resilient.

The functions of the skin

The skin is so much more than an external covering. It is a highly sensitive boundary between our bodies and the environment. The skin has several important functions, offering protection, temperature regulation and waste removal, as well as providing us with a sense of touch.

Protection

The skin acts as a protective organ in the following ways:

- The film of sebum and sweat on the surface of the skin, known as the acid mantle, acts as an antibacterial agent to help prevent the multiplication of micro-organisms on the skin.

- The fat cells in the subcutaneous layer of the skin help protect bones and major organs from injury.

- Melanin, which is produced in the basal cell layer of the skin, helps to protect the body from the harmful effects of ultraviolet radiation.

- The cells in the horny layer of the skin overlap like scales to prevent micro-organisms from penetrating the skin and to prevent excessive water loss from the body.

Temperature regulation

The skin helps to regulate body temperature in the following ways:

- When the body is losing too much heat, the blood capillaries near the skin surface contract to keep warm blood away from the surface of the skin and closer to major organs.

- The erector pili muscles raise the hairs and trap air next to the skin when heat needs to be retained.

- The adipose tissue in the dermis and the subcutaneous layer helps to insulate the body against heat loss.

- When the body is too warm, the blood capillaries dilate to allow warm blood to flow near to the surface of the skin, in order to cool the body.

- The evaporation of sweat from the surface of the skin will also assist in cooling the body.

Sensitivity

The skin is considered an extension of the nervous system. It is very sensitive to various stimuli due to its many sensory nerve endings which can detect changes in temperature and pressure and register pain.

Excretion

The skin functions as a miniature excretory system, eliminating waste through perspiration. The eccrine glands of the skin produce sweat, which helps to remove some waste materials from the skin such as urea, uric acid, ammonia and lactic acid.

Storage

The skin also acts as a storage depot for fat and water. About 15 per cent of the body's fluids are stored in the subcutaneous layer.

Absorption

The skin has limited absorption properties. Substances which can be absorbed by the epidermis include fat-soluble substances such as oxygen, carbon dioxide, fat-soluble vitamins and steroids, along with small amounts of water.

Key note

The skin is capable of absorbing small particles of substances such as essential oils due to the fact that they contain fat and water-soluble particles.

Vitamin D production

The skin synthesises vitamin D when exposed to ultraviolet light. Modified cholesterol molecules in the skin are converted by the ultraviolet rays in sunlight to vitamin D. It is then absorbed by the body for the maintenance of bones and the absorption of calcium and phosphorus in the diet.

The structure of the skin

Before looking at the structure of the skin, let's consider a few facts:

- The skin is a very large organ covering the whole body.
- It varies in thickness on different parts of the body. It is thinnest on the lips and eyelids, which must be light and flexible, and thickest on the soles of the feet and palms of the hands, where friction is needed for gripping.
- As the skin is the external covering of the body, it can be easily irritated and damaged and certain symptoms of disease and disorders may occur.
- Each person's skin varies in colour, texture and sensitivity and it is these individual characteristics that make each person unique.

The appearance of the skin reflects a person's physiology. Observation of a person's skin will indicate their nutrition, circulation, age, immunity, genetics as well as environmental factors which all play a significant role in the skin's colour, condition and tone.

Let's take a closer look at the structure of the skin. There are two main layers of the skin:

- the **epidermis**, which is the outer, thinner layer
- the **dermis**, which is the inner, thicker layer.

Below the dermis is the **subcutaneous layer** which attaches to underlying organs and tissues.

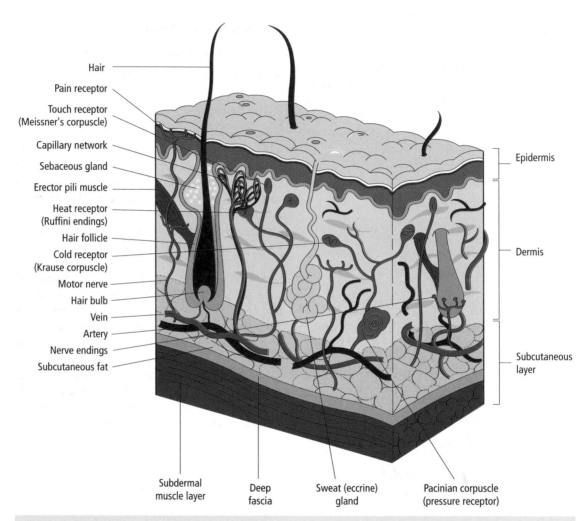

Hair

Pain receptor

Touch receptor
(Meissner's corpuscle)

Capillary network

Sebaceous gland

Erector pili muscle

Heat receptor
(Ruffini endings)

Hair follicle

Cold receptor
(Krause corpuscle)

Motor nerve

Hair bulb

Vein

Artery

Nerve endings

Subcutaneous fat

Epidermis

Dermis

Subcutaneous
layer

Subdermal
muscle layer

Deep
fascia

Sweat (eccrine)
gland

Pacinian corpuscle
(pressure receptor)

Figure 2.1   Structure of the skin

Key note

Keratin is the tough, fibrous protein found in the epidermis, hair and nails. The keratin found in the skin is constantly being shed. **Keratinisation** refers to the process that cells undergo when they change from living cells with a nucleus, which is essential for growth and reproduction, to dead, horny cells without a nucleus. Cells which have undergone keratinisation are therefore dead. **Keratin** has a protective function in the skin as the keratinised cells form a waterproof covering, helping to stop the penetration of bacteria and to protect the body from minor injury.

The epidermis

The epidermis is the most superficial layer of the skin and offers the body a waterproof, protective covering. It consists of five layers of cells. The three outermost layers consist of dead cells as a result of the process of **keratinisation**. The cells in the outermost layer are dead and scaly and are constantly being rubbed away by friction.

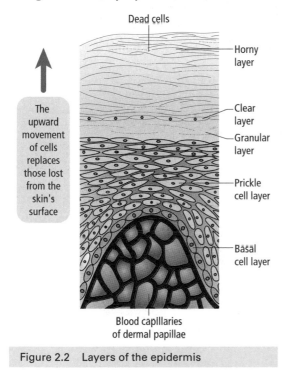

The upward movement of cells replaces those lost from the skin's surface

Dead cells
Horny layer
Clear layer
Granular layer
Prickle cell layer
Basal cell layer
Blood caplllaries of dermal papillae

Figure 2.2   Layers of the epidermis

The inner two layers are composed of living cells. The epidermis does not have a system of blood vessels, only a few nerve endings which are present in the lower epidermis. Therefore, all nutrients pass to the cells in the epidermis from blood vessels in the deeper dermis. The five layers of cells of the epidermis from the deepest to the most superficial are as follows:

- **stratum germinativum** (basal cell layer)
- **stratum spinosum** (prickle-cell layer)
- **stratum granulosum** (granular layer)
- **stratum lucidum** (clear layer)
- **stratum corneum** (horny layer).

Stratum germinativum (basal cell layer)

This is the deepest and innermost of the five layers. It consists of a single layer of column cells on a basement membrane which separates the epidermis from the dermis. In this layer the new epidermal cells are constantly being reproduced and therefore undergo continuous cell division.

These cells last about six weeks from reproduction or mitosis before being discarded into the horny layer. New cells are therefore formed by division, pushing adjacent cells towards the skin's surface. At intervals between the column cells, which divide to reproduce, are the large, star-shaped cells called melanocytes, which form the pigment melanin, the skin's main colouring agent. This layer also contains tactile (Merkel) discs that are sensitive to touch.

Key note

Melanin is produced by special cells called melanocytes which are found in the basal cell layer of the epidermis. Melanocytes have fingerlike projections which are capable of injecting melanin granules into neighbouring cells of the epidermis. This explains why melanin is found in the basal layer, prickle-cell layer and granular layer of the epidermis. Melanin is responsible for the colour of the skin and hair and helps protect the deeper layers of the skin from the damaging effects of ultraviolet radiation.

Stratum spinosum (prickle-cell layer)

This is known as the prickle-cell layer because each of the rounded cells contained within it has short projections which make contact with the neighbouring cells and give them a prickly appearance.

The stratum spinosum is a binding and transitional layer between the stratum granulosum and the stratum germinativum because it contains cells of both of these layers. The living cells of this layer are therefore capable of dividing by the process mitosis, but also assist in keratin formation. This layer also includes Langerhans cells which set up an immune response to foreign bodies.

Stratum granulosum (granular layer)

This layer consists of distinctly shaped cells, containing a number of granules which are involved in the hardening of the cells by the process keratinisation. The presence of keratin granules and disintegration of the cell nuclei are signs that the living cells have been transformed into a tough fibrous protein called keratin. This layer links the living cells of the epidermis to the dead cells above.

Stratum lucidum (clear layer)

This layer consists of small, tightly packed, transparent cells which permit light to pass through. It is thought to be the barrier zone controlling the transmission of water through the skin. It consists of three or four rows of flat, dead cells which are completely filled with keratin. They have no nuclei as the cells have undergone mitosis. The clear layer is very shallow in facial skin, but thick on the soles of the feet and palms of the hands, and is generally absent from hairy skin. This layer may also be absent from some thin skin.

Stratum corneum (horny layer)

This is the most superficial, outer layer, consisting of dead, flattened, keratinised cells which have taken approximately a month to travel from the germinating layer. This outer layer of dead cells is continually being shed by a process known as desquamation and replaced by cells from the deeper layers. The stratum corneum serves as an effective barrier against light and heat waves, bacteria and chemicals.

Key note

Cell regeneration occurs in the epidermis by the process mitosis. It takes approximately a month for a new cell to complete its journey from the basal cell layer, where it is reproduced, to the granular layer, where it becomes keratinised, to the horny layer where it is desquamated.

The dermis

The dermis is the deeper layer of the skin, and its key functions are to provide support, strength and elasticity. It is composed of dense connective tissue that is tough, extensible and elastic. The dermis has a higher water content than any other part of the skin and therefore has a key role in providing nourishment to the skin. The dermis has a superficial **papillary** layer and a deep **reticular** layer.

The papillary layer

The superficial **papillary** layer consists of areolar connective tissue and is connected to the underside of the epidermis by cone-shaped projections called **dermal papillae** which contain **nerve endings** (touch, pain, cold and heat) and a network of **blood** and **lymphatic** capillaries. The fine capillaries network in this layer brings oxygen and nutrients to the skin and carries the waste away. The many **dermal papillae** of the **papillary** layer form indentations in the overlying epidermis, giving it an irregular or ridged appearance. It is these ridges that leave fingerprints on objects that are handled. The key function of the papillary layer of the dermis is to provide vital nourishment to the living layers of the epidermis above.

The reticular layer

The deep **reticular** layer is situated below the papillary layer and is formed of dense, fibrous, connective tissue which contains the following:

- collagen fibres containing the protein **collagen** which gives the skin its strength and resilience

- elastic fibres containing a protein called **elastin** which gives the skin its elasticity

- **reticular** fibres which help to support and hold all structures in place.

The fibres in the **reticular** layer all help maintain the skin's tone, strength and elasticity.

Space between the fibres is occupied by a small quantity of adipose tissue, hair follicles, nerves, sweat glands and sebaceous glands. Cells present in the reticular layer of the dermis are as follows:

- **mast cells** which secrete histamine, causing dilation of blood vessels to bring blood to the area (this occurs when the skin is damaged or during an allergic reaction)

- **phagocytic cells** which are white blood cells that are able to travel around the dermis, destroying foreign matter and bacteria

- **fibroblasts** which are cells that form new fibrous tissue.

Key note

The principal function of the dermis is to provide nourishment to the epidermis and to give a supporting framework to the tissues.

Blood supply

Unlike the epidermis, the dermis has an abundant supply of blood vessels which run through the dermis and the subcutaneous layer of the skin. Arteries carry oxygenated blood to the skin via arterioles (small arteries) and these enter the dermis from below and branch into a network of capillaries around active or growing structures. These capillary networks form in the dermal papillae to provide the basal cell layer of the epidermis with food and oxygen. The networks also surround two appendages of the skin: the sweat glands and the erector pili muscles, which both have important functions in the skin.

The capillary networks drain into venules, small veins which carry the deoxygenated blood away from the skin and remove waste products. The dermis, therefore, is well supplied with capillary blood vessels to bring nutrients and oxygen to the germinating cells in the epidermis and to remove waste products from them.

Lymphatic vessels

The lymphatic vessels are numerous in the dermis and generally accompany the course of veins. They form a network through the dermis, allowing removal of waste from the skin's tissues. Lymph vessels are found around the dermal papillae, glands and hair follicles.

Nerves

Nerves are widely distributed throughout the dermis. Most nerves in the skin are sensory nerves which send signals to the brain and are sensitive to heat, cold, pain, pressure and touch. Branched nerve endings, which lie in the papillary layer and hair root, respond to touch and temperature changes. Nerve endings in the dermal papillae are sensitive to gentle pressure and those in the reticular layer are responsive to deep pressure.

Sensory nerves

There are at least five different types of sensory nerve endings in the skin:

- pain
- touch
- temperature (heat and cold)
- pressure.

The sensory nerve endings are also called receptors because they are part of the nervous system at which information is received.

Touch receptors

These receptors are located immediately below the epidermis. They are stimulated by light pressure on the skin, enabling a person to distinguish between different textures such as rough, smooth, hard and soft.

Pressure receptors

These receptors are situated beneath the dermis and stimulated by heavy pressure.

Pain receptors

These receptors consist of branched nerve endings in the epidermis and dermis. They are quite evenly distributed throughout the skin and are important in that they provide a warning signal of damage or injury in the body.

Temperature receptors

There are separate hot and cold receptors in the skin that are stimulated by sudden changes in temperature. The dermis also has motor nerve endings which relay impulses from the brain and are responsible for the dilation and constriction of blood vessels, the secretion of perspiration from the sweat glands and the contraction of the erector pili muscles attached to hair follicles.

The subcutaneous layer

This is a thick layer of connective tissue found below the dermis. The tissues areolar and adipose are present in this layer to help support delicate structures such as blood vessels and nerve endings. The subcutaneous layer contains the same collagen and elastin fibres as the dermis and also contains the major arteries and veins which supply the skin, forming a network throughout the dermis. The fat cells contained within this layer help to insulate the body by reducing heat loss. Below the **subcutaneous** layer of the skin lies the **subdermal muscle layer**.

Appendages of the skin

The appendages are accessory structures that lie in the dermis of the skin and project onto the surface through the epidermis. These include the **hair**, **erector pili muscle**, **sweat** and **sebaceous glands** and **nail**.

Hair is an important appendage of the skin which grows from a saclike depression in the epidermis called a hair follicle. Hair grows all over the body, with the exception of the palms of the hands and the soles of the feet, and is a sexual characteristic.

One of the primary functions of hair is physical protection. The eyelashes act as a line of defence by preventing the entry of foreign particles into the eyes and helping shade the eyes from the sun's rays. Eyebrow hairs help to divert water and other chemical substances away from the eyes. Hairs lining the ears and the nose trap dust and help to prevent bacteria from entering the body. Body hair acts as a protective barrier against the sun and helps to protect us against the cold, with the help of the erector pili muscle.

Another function is preventing friction. Underarm and pubic hair protects the skin and cushions against friction caused by movement.

The structure of a hair

The hair is composed mainly of the protein keratin and is therefore a dead structure. Longitudinally, the hair is divided into three parts:

- **hair shaft** – the part of the hair lying above the surface of the skin

- **hair root** – the part found below the surface of the skin

- **hair bulb** – the enlarged part at the base of the hair root.

Internally, the hair has three layers which all develop from the matrix (the active growing part of the hair):

- **cuticle** – the outer layer made up of transparent protective scales which overlap one another. The cuticle protects the cortex and gives the hair its elasticity.

- **cortex** – the middle layer made up of tightly packed keratinised cells containing the pigment melanin which gives the hair its colour. The cortex helps to give strength to the hair.

- **medulla** – the inner layer made up of loosely connected keratinised cells and tiny air spaces. This layer of the hair determines the sheen and colour of hair due to the reflection of light through the air spaces.

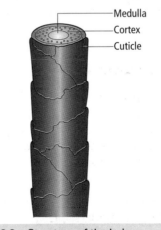

Medulla
Cortex
Cuticle

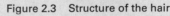

Figure 2.3   Structure of the hair

Hair colour is due to the presence of melanin in the cortex and medulla of the hair shaft. In addition to the standard black colour, the melanocytes in the hair bulb produce two colour variations of melanin: brown and yellow. Blond, light-coloured and red hair has a high proportion of the yellow variant. Brown and black hair possesses more of the brown and black melanin.

The structure of a hair in its follicle

The individual parts of a hair's structure are as follows:

Connective tissue sheath

The connective tissue sheath surrounds both the follicle and sebaceous gland. Its function is to supply the follicle with nerves and blood. It is the main source of sustenance for the follicle.

Outer root sheath

The outer root sheath forms the follicle wall and is continuous with the basal cell layer of the epidermis. It provides a permanent source of growing cells (hair germ cells) to enable the follicle to grow and renew cells during its life cycle.

Dermal papilla

The dermal papilla is an elevation at the base of the bulb which contains a rich blood supply. It serves as a crucial source of nourishment for hair, providing the hair cells with food and oxygen.

Inner root sheath

The inner root sheath originates from the dermal papilla at the base of the follicle and grows upwards with the hair. The inner root sheath is made up of similar cells to the cuticle of the hair. They lie in the opposite direction,

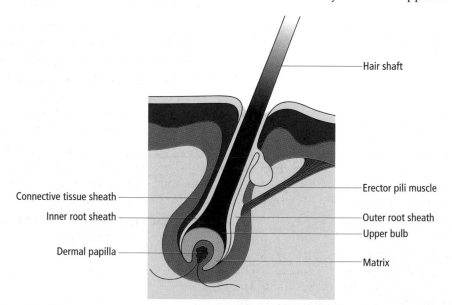

Figure 2.4   A hair in its follicle

facing downwards towards the dermal papilla. This allows the cells of the inner and outer layer to lock together, thereby shaping and contouring the hair, helping to anchor it into the follicle. The inner root sheath ceases to grow when level with the sebaceous gland.

Hair bulb

The hair bulb is the enlarged part at the base of the hair root. A gap at the base leads to a cavity which contains the dermal papilla. The hair bulb is where the cells grow and divide by the process of mitosis.

Matrix

The matrix is located at the lower part of the hair bulb. It is the area of mitotic activity of the hair cells.

Types of hair

There are three main types of hair in the body: **lanugo**, **vellus** and **terminal** hair.

Lanugo hair

Lanugo is the fine, soft hair found on a foetus. It grows from around the third to the fifth month of pregnancy and is eventually shed to be replaced by secondary vellus hairs, around the seventh to the eight month of the pregnancy. Lanugo hair is often unpigmented and, as it is soft and fine, lacks a medulla.

Vellus hair

Vellus hair is the soft, downy hair found all over the face and body, except for the palms of the hands, soles of the feet, eyelids and lips. Vellus hairs are often unpigmented and do not have a medulla or a well-developed bulb. Vellus hair lies close to the surface of the skin and, therefore, has a shallow follicle. If stimulated by an increase in blood circulation resulting from hormonal changes in the body (such as puberty, pregnancy or menopause) or medication, the shallow follicle of a vellus hair can grow downwards to become a coarse, dark, terminal hair.

Terminal hair

Terminal hairs are the longer, coarser hairs found on the scalp, under the arms, eyebrows, pubic regions, arms and legs, and most are pigmented. They vary greatly in shape, diameter, length, colour and texture. They are deeply seated in the dermis and have well-defined bulbs.

Facts about hair growth

- Hair begins to form in the foetus from the third month of pregnancy.
- The growth of hair originates from the matrix, which is the active growing area where cells divide and reproduce by mitosis.
- Living cells, which are produced in the matrix, are pushed upwards away from their source of nutrition, die and are converted to keratin to produce a hair.
- Hair has a growth pattern which ranges from approximately four to five months for an eyelash hair to approximately four to seven years for a scalp hair.
- Hair growth is affected by illness, diet and hormonal influences.

The growth cycle of a hair

Each hair has its own growth cycle and undergoes three distinct stages of development: **anagen**, **catagen** and **telogen**.

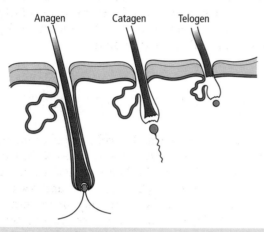

Figure 2.5   The hair growth cycle

Anagen

- active growing stage
- lasts from a few months to several years
- hair germ cells reproduce at matrix
- new follicle is produced which extends in depth and width
- the hair cells pass upwards to form the hair bulb
- hair cells continue rising up the follicle, and as they pass through the bulb they differentiate to form individual structures of hair
- inner root sheath grows up with the hair, anchoring it into the follicle

- when cells reach the upper part of the bulb they become keratinised
- two-thirds of its way up the follicle, the hair leaves an inner root sheath and emerges on to the surface of the skin.

Catagen

- lasts approximately two to four weeks
- transitional stage from active to resting
- hair separates from dermal papilla and moves slowly up the follicle
- follicle below retreating hair shrinks
- hair rises to just below level of sebaceous gland where the inner root sheath dissolves and the hair can be brushed out.

Telogen

- short resting stage
- shortened follicle rests until stimulated once more
- hair is shed on to the skin's surface
- new replacement hair begins to grow.

Different types of hair growth

As there is a continuous cycle of hair growth, the amount of hair on the body remains fairly constant. However, hair growth will vary from person to person and from area to area. A new client coming to the salon for a hair removal treatment should be made aware of the fact that hair growth occurs in three stages which will result in the hair being at different lengths both above the skin and below it.

Key note

While carrying out hair removal treatments it is important to remember that the hair follicle is part of the skin's structure; therefore any treatment which affects the hair is also going to affect the skin. Once a hair has been removed, the maximum amount of blood will be sent straight to the area being treated to heal and protect the skin. This is a normal reaction of the skin and extra blood that has been sent to the treated area will be diverted again within a few hours of treatment. As the treated area of skin will have open follicles, it is vital that a client adheres strictly to aftercare advice specified, as open follicles offer bacteria an easy entry into the body.

Erector pili muscle

This is a small, smooth, weak muscle made up of sensory fibres. It is attached at an angle to the base of a hair follicle which serves to make the hair stand erect in response to cold, or when experiencing emotions such as fright and anxiety.

Sweat glands

There are two types of sweat glands in the skin: **eccrine** and **apocrine**. The majority are **eccrine** glands which are simple, coiled, tubular glands that open directly onto the surface of the skin. There are several million of them distributed over the surface of the skin, although they are most numerous on the palms of the hands and the soles of the feet.

Their function is to regulate body temperature and help eliminate waste products. Their active secretion of sweat is under the control of the sympathetic nervous system. Heat-induced sweating tends to begin on the forehead and then spreads to the rest of the body. Emotionally induced sweating, stimulated by fright, embarrassment or anxiety, begins on the palms of the hands and in the axillae then spreads to the rest of the body.

Apocrine glands are connected with hair follicles and are only found in the genital and underarm regions. They produce a fatty secretion; breakdown of the secretion by bacteria leads to body odour.

Sebaceous glands

These glands are small, saclike pouches found all over the body, except for the soles of the feet and the palms of the hands. They are more numerous on the scalp, face, chest and back.

Sebaceous glands commonly open into a hair follicle, but some open on to the skin surface. They produce an oily substance called **sebum** which contains fats, cholesterol and cellular debris.

Sebum is mildly antibacterial and anti-fungal and coats the surface of the skin and the hair shafts, where it prevents excess water loss, lubricates and softens the horny layer of the epidermis, and softens the hair. The secretion of sebum is stimulated by the release of hormones, primarily androgens.

The nail

The nail is an important appendage of the skin and is an extension of the stratum lucidum (clear layer) of the epidermis. It is composed of horny, flattened cells which undergo a process of keratinisation, giving the nail a hard appearance. It is the protein keratin which helps to make the nail a strong but flexible structure. The part of the nail the eye can see is dead as it has no direct supply of blood, lymph and nerves. All nutrients are supplied to the nail via the dermis.

Functions of the nail

The nail has several important functions:

- It forms a protective covering at the ends of the phalangeal joints of the fingers and the toes, helping to support the delicate network of blood vessels and nerves at the ends of the fingers.

- It is a useful tool, enabling us to touch, manipulate small objects and scratch surfaces.

The structure of the nail

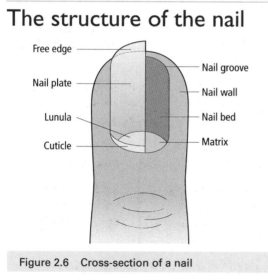

Figure 2.6   Cross-section of a nail

The nail has several important anatomical regions.

Nail matrix

The matrix is situated immediately below the cuticle and is the nail's most important feature. It is the area where the living cells are produced. The matrix receives a rich supply of blood which supplies oxygen to the nail and is vital to the production of new cells. It is the area from which the health of the nail is determined.

Nail bed

The nail bed is situated immediately below the nail plate and is a continuation of the matrix. It is the part of the skin upon which the nail plate rests. The nail bed is richly supplied with blood vessels, lymph vessels and nerves from the underlying dermis. The key functions of the nail bed are to provide nourishment and protection for the nail.

Cuticle

The cuticle is a fold of overlapping skin that surrounds the base of the nail. There are different names given to the different areas of the cuticle:

- the eponychium is the dead cuticle that adheres to the base of the nail, near the lanula

- the peronychium is the cuticle that outlines the nail plate

- the hyponychium is the cuticle skin found under the free edge of the nail.

The key function of the cuticle is to protect the matrix and provide a protective seal against bacteria.

Lunula

The lunula is the light-coloured, semicircular area of the nail, commonly called the half-moon, that lies in between the matrix and the nail plate. The lunula is always present but not always visible, as it may be obscured by the cuticle. It is an area of the nail where cells start to harden; the cells here are in a transitional stage (between hard and soft). The lunula is therefore a bridge between the living cells of the matrix and the dead cells of the nail plate.

Nail plate

The nail plate is the main visible part of the nail which rests on the nail bed and ends at the free edge. It is made up of layers of translucent, dead, keratinised cells to make the nail hard and strong. The layers of cells are packed very closely together with fat but very little moisture. The key function of the nail plate is to offer protection for the nail bed.

Nail walls

The nail walls are the folds of skin overlapping the sides of the nails. They surround three sides of the nail and are firmly attached to the sides of the nail plate. The key function of the nail walls is to protect the edges of the nail plate from external damage.

Nail grooves

The nail grooves are the deep ridges under the sides of the nail. As the nail grows along with the nail bed, it passes along the nail grooves which guide it and help it to grow straight.

Free edge

The free edge is the part of the nail plate that extends beyond the nail bed. It is the part of the nail that is filed and is usually the hardest part.

Facts about nail growth

- Nails start growing on a foetus before the fourth month of pregnancy.
- Nail growth occurs from the nail matrix by cell division.
- As new cells are produced in the matrix, older cells are pushed forward and hardened by the process of keratinisation which forms the hardened nail plate.
- As the nail grows it moves along the nail grooves at the sides of the nail, which helps to direct the nail growth along the nail bed.
- It takes approximately six months for cells to travel from the lunula to the free edge of the nail.
- The growth of a nail does not follow a growth cycle and hence growth is continuous throughout life.
- The average growth rate of a nail is approximately 3 mm per month.
- The growth rate of nails will vary from person to person and from finger to finger, with the index finger generally being the fastest to grow.
- Toenails have a slower rate of growth than fingernails.
- The rate of growth of a nail is faster in the summer due to an increase in cell division as a result of exposure to ultraviolet radiation.
- A good blood supply is essential to nail growth; oxygen and nutrients are fed to the living cells of the nail matrix and nail bed.

- Protein and calcium are good sources of nourishment for the nails.

Nail growth may be affected by the following factors:

- Ill health – during illness the body receives a reduced blood supply to the nails as it attempts to restore the rest of the body to good health.

- Diet – a nutritional deficiency can result in a diminished blood supply to the nail.

- Age – during ageing the growth of a nail slows down due to the fact that the blood vessels supplying the matrix and the nail become less efficient.

- Poor technique – if heavy pressure is used when using manicure implements such as a cuticle knife, damage may be caused to the matrix cells, resulting in ridges to the nail. This may only be temporary as new cells produced in the matrix will replace the damaged ones, and depending on the extent of the damage the ridges may eventually grow out.

- An accident – such as shutting a finger in the door. This may result in bruising and bleeding of the nail or even the complete removal of a nail. It could result in permanent malformation of the nail if the nail bed has become damaged.

Nail diseases and nail disorders

Diseases of the nail are as a direct result of bacteria, fungi, parasites or viruses attacking the nail or surrounding tissues. Nail disorders may be caused by illness, physical and chemical damage, by general neglect or by poor manicuring techniques. Nail disorders do not contraindicate manicure or pedicure treatments. However, nail diseases do as they may cause cross-infection.

A therapist must be able to recognise diseases and disorders so that the correct treatment or advice may be given.

Common nail diseases

Paronychia

Inflammation of the skin surrounding the nail. The tissues may be swollen and pus may be present which can develop into an abscess. It is a common condition on the fingers and is caused by bacterial or viral infection.

Initially, the cause may be due to prolonged immersion of the hands in water, poor manicure techniques, picking the cuticle or the nail wall separating from the nail. Infection with the herpes simplex virus can give rise to a whitlow, an abscess that forms around the nail.

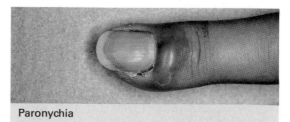

Paronychia

Onychomycosis

This is a term given to fungal infections of the nail, commonly called ringworm. It attacks the nail bed and nail plate, and presents as white or yellow scaly deposits at the free edge, which may spread down to invade the nail walls or bed. The nails become thickened, brittle, opaque or discoloured. The nail plate will appear spongy and furrowed.

In its advanced stages, the nail plate may separate from the nail bed (a condition known as onycholysis (see below). There may also be accompanying dryness and skin scaling at the base of the fingers and on the palms.

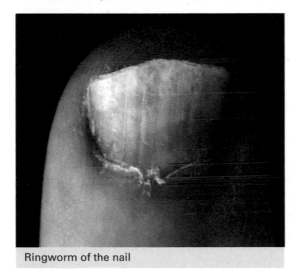

Ringworm of the nail

Onychia

This is a generic term used to describe any disease of the nail, but more specifically refers to inflammation of the nail bed. In this condition the nail matrix appears red. There may be swelling, tenderness and pus formation. This could lead to the nail being shed. This condition may be caused by wearing false nails for too long or by harsh manicuring, chemical applications or by a variety of infections or physical damage.

Onycholysis

This is separation or loosening of part or all of a nail from its bed. It may be due to disease, physical damage or may occur spontaneously without any apparent cause. It can occur if sharp instruments are used under the free edge. Penetration of the flesh line allows bacteria or other infection to enter the nail bed.

Nail disorders

Leuconychia

This is a term given to white or colourless nails, or nails with white spots, streaks or bands. There may also be evidence of ridging. It may be caused as a result of injury to the matrix or the effects of disease. The white spots will usually disappear as the nail grows.

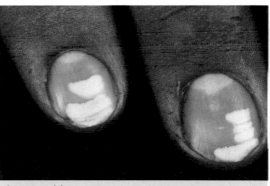

Leuconychia

Onychophagy

This is the technical term for nail biting in which the free edge, nail plate and cuticle are bitten to leave the hyponychium exposed and the cuticle and surrounding skin ragged, inflamed and sore. Nail biting is usually a nervous or stress-induced habit.

Nail ridges/corrugations

Ridges in the nail may occur due to irregular formation of the nail or to physical/chemical injury of the nail matrix. Ridges may be vertical which are common in healthy nails due to uneven development of the nail tissue, poor manicuring techniques or the effects of harsh chemicals. Ridges may also be horizontal and can be indicative of abnormal nail growth, a symptom of body malfunction or disease. Deep horizontal lines are often associated with illness.

Hangnail

A hangnail is a small strip of skin that hangs loosely at the side of the nail, or a small portion of the nail itself splitting away. A hangnail may develop due to dry, torn or split cuticles. Common causes are hands being immersed in water for long periods, cutting the nails too close, digging the cuticles, improper filing or the effects of detergents and other chemicals.

Onychogryphosis

This is a term given to an ingrown fingernail or toenail. The first signs are inflammation, followed by tenderness, swelling and pain. Infection may aggravate the condition. It is caused by ill-fitting shoes, cutting or filing nails too short or too close to the skin. It may also be due to a malformation of the nail when it was beginning to grow.

Pterygium

This is a condition where the cuticle becomes overgrown and excessive and grows forward. The cuticle at the base of the nail becomes dry and split and grows forward, sticking to the nail plate. Pterygium may be due to faulty nail care or lack of nail care.

Koilonychia

This is the term given to concave, spoon-shaped nails. In this condition the nails are thin, soft and hollowed. Koilonychia may be congenital or it may be due to lack of iron or other minerals. The spoon-shape results from abnormal growth at the nail matrix.

Eggshell nails

This is a term given to thin, white nails that are more flexible than normal. In this condition the nail separates from the nail bed and curves at the free edges. The condition may be associated with illness.

Onychorrhexis

This is the term given to dry, brittle nails. In this condition the nail loses its moisture, becomes dry and the free edge splits. The nails may peel into layers very easily. There may be transverse, or longitudinal splitting of the nail plate, and inflammation, tenderness, pain, swelling and infection may be present. Frequent immersion in water and contact with detergents and chemicals contribute to this condition. It may also indicate an iron deficiency, anaemia or incorrect filing which causes the nail plate to split.

Factors affecting the skin

Diet

A healthy body is needed for a healthy skin. The skin can be thought of as a barometer of the body's general health. A nutritionally balanced diet is vital to the skin, as it will contain all the essential nutrients to nourish the cells in growth and repair. In addition to the essential nutrients in the diet, the following vitamins aid the regeneration of the skin:

- Vitamin A – helps repair the body's tissues and helps prevent dryness and ageing
- Vitamin B – helps improve the circulation and the skin's colour, and is essential for cellular oxidation
- Vitamin C – is essential for healing and to maintain levels of collagen in the skin.

Water

Drinking an adequate amount of water (approximately six to eight glasses per day) aids the digestive system and helps to prevent a build-up of toxicity in the tissues of the skin.

Sleep

Sleep is essential to physical and emotional well-being and is one of the most effective regenerators for the skin.

Stress and tension

When the body is subjected to regular stress and tension it can cause sensitivity and allergies in the skin, as well as encourage the formation of lines around the eyes and the mouth.

Exercise

Regular exercise promotes good circulation, increased oxygen intake and blood flow to the skin.

Alcohol

Alcohol has a dehydrating effect on the skin and excess consumption causes the blood vessels in the skin to dilate.

Smoking

Smoking affects the skin's cells and destroys vitamins B and C which are important for a healthy skin. Smoking dulls the skin by polluting the pores and increases the formation of lines around the eyes and the mouth.

Medication

Medication can affect the skin by causing dehydration, oedema, sensitivity and/or allergies.

Ultraviolet light

The action of ultraviolet light (UVA), whether from a natural or artificial source can be hazardous to the skin. UVA penetrates deep into the dermis where it can cause premature ageing of the skin. Highly reactive molecules called free radicals are formed which cause skin cells to degenerate and lose their elasticity.

UVB can cause sunburn, in which the cells become red and damaged and the skin may blister. UVB is also implicated in types of skin cancers, particularly malignant melanoma.

Chemicals

Harsh alkaline chemicals in products containing detergents and soaps can cause moisture loss in the skin by stripping sebum from the skin's surface.

Climate

The climate has several effects on the skin:

- A cold climate can lead to a reduced production of sebum, resulting in reduced protection for the skin and increased water evaporation.

- A hot climate can lead to increased moisture loss through perspiration.

- Extremes of temperature (from hot to cold or cold to hot) can lead to the formation of broken capillaries).

Environment

Air pollution such as carbon from smoke, fumes from car exhausts, chemicals from factories and the environment, can all be potentially damaging to the skin's cells. Environmental pollutants such as lead, mercury and aluminium can all accumulate in the body and may find their way into food through polluted waters, rain and dust.

Hormones

The natural glandular changes of the body have an effect on the condition of the skin throughout life:

- During puberty the sex hormones stimulate the sebaceous glands which may cause some imbalance in the skin.

- At the onset of menstruation the skin may erupt due to the adjustment of hormone levels at that time.

- During pregnancy pigmentation changes may occur, but usually disappear at birth.

- During the menopause the activity of the sebaceous glands is reduced and the skin becomes drier.

Age

The natural process of ageing naturally affects the skin. From the mid-thirties, the skin starts to lose its firmness and fine lines and wrinkles start to appear. In the forties and fifties, lines and wrinkles will deepen and loss of muscle tone causes sagging of the skin on the cheeks and the neck. The connective tissue in the skin loses its elasticity and becomes less firm, and the skin becomes thinner and finer. As part of the ageing process, cell regeneration in the skin decreases and the skin appears dry and dull.

Skin types and their characteristics

Skin is normally classified into four main types:

- **normal/balanced**
- **greasy/oily**
- **dry**
- **combination**.

Additional characteristics may or may not be present so that the skin is further described as:

- **sensitive**
- **congested**
- **dehydrated**
- **blemished**
- **mature**.

Normal/balanced skin

This is the perfect skin which is very rare indeed. Usually only a younger skin is 'normal'. In a balanced skin, as the name suggests, there exists a **correct balance** of **oil/sebum** and **moisture**.

Distinguishing features

- The skin is neither too dry nor too oily.
- The skin should be soft and supple to the touch.
- A smooth texture which is neither too thick nor too thin.
- The skin should feel slightly warm, denoting a good blood supply.
- The skin should have an even surface free from blemishes. If the odd blemish forms they should heal easily and quickly.
- The pores should be fine and virtually non-apparent.
- The skin should feel firm to the touch and generally have good elasticity.

Dry skin

A dry skin is so-called because it is either lacking in sebum or moisture or both. The skin's natural oil sebum lubricates the corneum layer, and in the absence of this oily coating the dead cells start to curl up and flake. The sebum coating also helps to prevent moisture loss by evaporation.

Distinguishing features

- The skin looks dry and often parched.
- The skin feels papery and even a little coarse to the touch.
- The texture appears thin and coarse and there may be patches of flaking skin.
- The skin looks like 'parchment' and is often sensitive and prone to the formation of dilated capillaries and milia around the eye and upper cheek area.
- Dry skin tends to age prematurely, with fine lines becoming evident around the eyes even as early as in the early or mid-twenties.
- The pores are small and tight due to the lack of sebum production.
- The skin does not usually have good elasticity.

Oily/greasy skin

An oily skin produces too much sebum. Usually this is common at puberty when the sebaceous glands are stimulated by the male hormone androgen, which is produced in both males and females but in larger quantities in the male. The activity of the sebaceous glands usually decreases in the twenties.

Distinguishing features

- The skin appears oily, a characteristic shine is often apparent, especially down the T-zone.

- The skin feels thick and coarse to the touch.

- The texture is usually uneven.

- The skin is sallow in colour as a result of excess sebum production and dead corneum cells being permitted to build up on the outer surface.

- Blemishes are often very apparent, with blocked pores, comedones, papules and pustules all being present to a greater or lesser degree. Often some scarring is evident from previous blemish sites which lead to a very uneven surface colour.

- The pores are large and noticeable due to a build-up of sebum causing them to stretch open.

- The skin usually feels firm to the touch and an oily skin ages least prematurely.

- Elasticity is generally good.

Combination skin

As its name suggest this skin is a bit of a **mixture**. Typically the T-zone is oily and the cheeks and neck dry/normal. This is actually the **most common skin type**.

Distinguishing features

- The skin is dry on the cheeks and neck and oily on the T-zone.

- The skin will probably vary, with the dry areas feeling rough and fine and the oily areas feeling thicker and coarse to the touch.

- The skin will have a patchy colour and the T-zone will probably suffer from blemishes, such as blocked pores, comedones, papules and pustules.

- Milia may be present around the dryer skin areas, with some sensitivity and dilated capillaries also evident.

- The pores are fine and small on the cheeks and neck but larger in the T-zone.

- The skin's tone and elasticity varies, being poor in the dry areas but good in the oily areas.

Sensitive skin

This often accompanies a dry skin, due to its lack of protective sebum and low moisture content. The skin is basically hypersensitive and reacts even to the mildest stimulus. The skin flushes very easily, which when occurring frequently often results in the formation of dilated capillaries.

Distinguishing features

- There is high colouring with or without dilated capillaries.

- The skin is usually warm to the touch.

- Even after a gentle cleanse the skin may show quite high colouring.

Dehydrated skin

A dehydrated skin lacks moisture and this can affect any skin type, even an oily skin. If an oily skin is also dehydrated this usually results from using products that are too harsh and have left the skin stripped of its protective coating of sebum. Many skin types can suffer from temporary dehydration, caused through illness, medication, overexposure to the elements and central heating.

Distinguishing features

■ There is superficial flaking and lining, with a parched, dry-looking, rough surface.

■ Dilated capillaries may be present.

Mature skin

A mature skin is beginning to show signs of ageing. Although we begin to age the moment we are born, aesthetically we begin to age around the mid-twenties. The rate at which the ageing process progresses depends on many factors, the strongest being genetics and racial tendencies. Darker-skinned races age far less quickly than fair Caucasians. Next is exposure to UV light, which can be very damaging, incorrect skin care or lack of it, diet, illness and medication. The skin type also contributes, with oily skin ageing less quickly than dry skin. As a woman comes to the end of her reproductive years and reaches the menopause the most dramatic signs of ageing can be seen, due to a reduction in the production of female hormones.

Distinguishing features

■ Skin becomes dryer as the sweat and sebaceous glands become less active.

■ There is a loss of elasticity as the elastic fibres harden. When the skin is pulled with the fingers it does not 'spring back' quickly like a young skin.

■ Fine lining develops and wrinkling sets in as the collagen fibres lose their ability to hold water effectively and become cross-linked.

■ The skin begins to sag as muscles lose their tone.

■ The rate of cell renewal slows so that the skin becomes finer and more transparent and the blood supply becomes evident.

■ There is a reduction in supporting fat and the bony structure becomes more evident and the skin loses its youthful plumpness.

■ As the circulation slows the skin may become quite sallow in colour and also puffy.

■ There is often pigmentation on the skin and other typical ageing abnormalities present such as age spots, skin tags (common on the neck) and seborrhoeic warts.

■ Dilated capillaries are often present.

■ Superfluous hair growth developing on the upper lip and other areas of the face.

Ethnic skin types

All ethnic skin types vary in the degree of melanin they produce. Although all ethnic skin types have the same number of melanocyte cells, black skins have melanocytes capable of making large amounts of melanin.

White skin

The colour of white skin (British, Scandinavian, East and West European, North American, South Australian, Canadian, New Zealand origin) is generally a pale buff. There are relatively small amounts of melanin present

White skin

in white skins, as melanin is produced to varying degrees. Some white skins are pale and translucent, often accompanied by freckles and red, blond or mousy hair. Pale skin is at risk from sunburn, ageing and the formation of skin cancer because of the reduced protection it has from the lack of melanin in the skin. Pale skins may be unable to develop a tan and may be prone to dehydration and irritation.

Other types of white skins tan more easily and are far less sensitive, and while being pale in the winter may establish a golden tan easily without burning. Some skins may appear pinkish while others have a sallowish tone. White skins age faster than black skins and it is important, therefore, to start protecting the skin from ultraviolet radiation as early as possible.

Oriental/light Asian skin

Oriental/light Asian skin (Chinese, Japanese or Middle Eastern origin) is a creamy colour with a tendency to yellow and olive tones, with more melanin present. Oriental skin rarely shows blemishes and defies normal signs of ageing. Scars are more likely to occur and hyperpigment, causing unevenness, troughs, pits and hollows on the skin's surface.

Oriental/light Asian skin

Dark Asian skin

Dark Asian skin (Pakistani, Indian, Sri Lankan or Malaysian origin) is a very dark skin colour which is deeply pigmented with melanin. Dark Asian skin is smooth and supple, with minimal signs of ageing. Sweat glands are larger and more numerous in this skin type, which gives a sheen to the skin that is often mistaken for oiliness. As this skin type is deeply pigmented it does not reveal the blood capillaries.

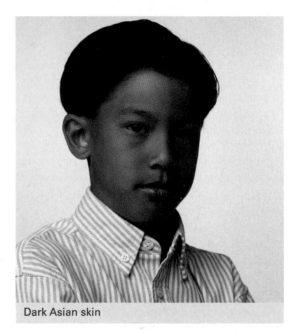
Dark Asian skin

Mediterranean skin

Mediterranean skin (Italian, Spanish, Greek, Portuguese, Yugoslavian, South American or Central American origin) looks sallow, with some reddish pigment. There is a good degree of melanin present which obscures the colour of the blood vessels. This skin type tends to have a generous coating of sebum and is therefore oily. This skin type tans easily and deeply without burning.

Mediterranean skin

Dermatosis papulosa nigra (DPN)

Afro-Caribbean/black skin

Black skins (West Indian/African origin) have a higher degree of sebum as sebaceous glands are larger, more numerous and closer to the skin's surface with open pores. This skin type is thick and tough, desquamates easily and forms keloid scars when damaged (keloid scars are thick and lumpy and often extend beyond the borders of the wound). It is also more likely to be affected by several different types of disfiguring bumps. **Dermatosis papulosa nigra** (DPN) is a benign cutaneous condition that is common in black skins. It is characterised by multiple, small, hyperpigmented, asymptomatic papules.

As black skin is thicker than white skin it is prone to congestion and comedones. Black skin generally ages at a much slower rate than white skin, mainly due to the extra protection afforded by the melanin. A disadvantage to having more melanin is that it makes the skin more reactive. This means that almost any stimulus, such as a rash, scratch or inflammation, may trigger the production of excess melanin, resulting in dark marks or patches on the skin. This is known as **post-inflammatory hyperpigmentation**.

Occasionally some black skins develop a decrease in melanin, or **post-inflammatory** hypopigmentation, in response to skin trauma. In either case (hypo- or hyperpigmentation) the light or dark areas may be disfiguring and may take months or years to fade. The increased thickness of the horny layer of the skin in black skins can cause dehydration which leads to increased skin shedding. This can create an ashen effect as the loose cells build up on the skin.

Afro-Caribbean/black skin

Mixed skin

Clients with a mixed skin will usually have a combination of characteristics of all the above skin types. The shades of colour and characteristics will vary greatly depending on the mix.

Male skin

Although there is not specifically a 'male' skin type, it is important to consider the differences in make-up between male and female skins. First there are hormonal differences between men and women. Testosterone, the male hormone, gives men a thicker epidermis (approximately 2 mm compared to 1.5 mm in women). Male skin does have a tendency to be tougher, more elastic and less sensitive than female skin, although daily shaving can increase the risk of skin rashes, infections and ingrowing hairs. It is also more acidic and has a more efficient supply of blood and sebum. This means it tends to age better than women's skin, remaining softer, firmer and suppler.

General terms associated with the skin

Allergic reaction

This disorder occurs when the body becomes hypersensitive to a particular allergen. When irritated by an allergen, the body produces histamine in the skin as part of the body's defence or immune system. The effects of different allergens are diverse and they affect different tissues and organs. For example, certain cosmetics and chemicals can cause rashes and irritation in the skin. Certain allergens, such as pollen, fur, feathers, mould and dust, can cause asthma and hay fever. If severe, allergies may be extremely serious and result in anaphylactic shock.

Comedone

This is a collection of sebum, keratinised cells and waste which accumulate in the entrance of a hair follicle. It may be open or closed. An open comedone is a blackhead contained within the follicle, whereas a closed comedone is a whitehead, trapped underneath the skin's surface.

Crow's feet

These are fine lines around the eyes caused by habitual facial expressions and daily movement. They are associated with ageing of muscle tissue, but premature formation may be due to overexposure to UV light or eye strain.

Cyst

This is an abnormal sac containing liquid or a semi-solid substance. Most cysts are harmless.

Erythema

This is reddening of the skin due to the dilation of blood capillaries just below the epidermis in the dermis.

Fissure

This is a crack in the epidermis exposing the dermis.

Keloid

A keloid is the overgrowth of an existing scar which grows much larger than the original wound. The surface may be smooth, shiny or ridged. The onset is gradual and is due to an accumulation or increase in collagen in the immediate area. The colour varies from red, fading to pink and white.

Lesion

A zone of tissue with impaired function, as a result of damage by disease or wounding, is called a lesion.

Macule

A macule is a small, flat patch of increased pigmentation or discolouration such as a freckle.

Milia

Milia is sebum trapped in a blind duct with no surface opening. Usually found around the eye area, they appear as pearly, white and hard nodules under the skin.

Mole

Moles are also known as a pigmented naevi. They appear as round, smooth lumps on the surface of the skin. They may be flat or raised and vary in size and colour from pink to brown or black. They may have hairs growing out of them.

Naevus

This is a mass of dilated capillaries and may be pigmented as in a birthmark.

Papule

A papule is a small, raised elevation on the skin, less than 1 cm in diameter, which may be a red colour. It often develops into a pustule.

Pustule

This is a small, raised elevation on the skin containing pus.

Skin tag

A small growth of fibrous tissue which stand up from the skin and sometimes is pigmented (black or brown).

Scar

A scar is a mark left on the skin after a wound has healed. Scars are formed from replacement tissue during the healing of a wound. Depending on the type and extent of damage, the scar may be raised (hypertrophic), rough and pitted (ice-pick) or fibrous and lumpy (keloid). Scar tissue may appear smooth and shiny or form a depression in the surface.

Telangiecstasis

This is the term for dilated capillaries, where there is persistent vasodilation of capillaries in the skin. It is usually caused by extremes of temperature and overstimulation of the tissues, although sensitive and fair skins are more susceptible to this condition.

Tumour

A tumour is formed by an overgrowth of cells. Almost every type of cell in the epidermis and dermis is capable of benign or malignant overgrowth. Tumours are lumpy, and even when they cannot be seen they can be felt underneath the surface of the skin.

Ulcer

An ulcer is a break or open sore in the skin extending to every layer.

Urticaria

This condition is also known as hives. Lesions appear rapidly and disappear within minutes or gradually over a number of hours. The clinical signs are the development of red weals which may later turn white. The area becomes itchy or may sting. There are a number of causes of urticaria, some of which are an allergic reaction to certain foods such as strawberries, shellfish, penicillin, house dust and pet fur. Other causes include stress and sensitivity to light, heat or cold.

Vesicles

These are small, saclike blisters. A bulla is a vesicle larger than 0.5 cm and is commonly called a blister.

Wart

A wart is a well-defined, benign tumour, varying in size and shape. (See viral infections, page 53).

Weal

A weal is a raised area of skin, containing fluid which is white in the centre with a red edge. It is seen in the condition urticaria.

Disorders of the sebaceous gland

Acne vulgaris

This is a common inflammatory disorder of the sebaceous glands which leads to the overproduction of sebum. It involves the face, back and chest and is characterised by the presence of comedones, papules and, in more severe cases, cysts and scars. Acne vulgaris is primarily androgen-induced. It appears most frequently at puberty and usually persists for a considerable period of time.

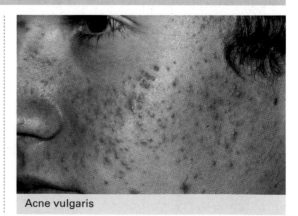

Acne vulgaris

Rosacea

A chronic inflammatory disease of the face in which the skin appears abnormally red. The condition is gradual, and begins with flushing of the cheeks and nose. As the condition progresses it can become pustular. Aggravating factors include hot, spicy foods, hot drinks, alcohol, menopause, the weather and stress.

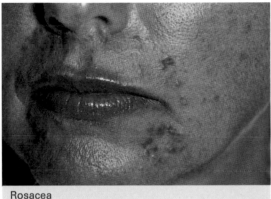

Rosacea

Sebaceous cyst

This type of cyst is a round, nodular lesion with a smooth, shiny surface which develops from a sebaceous gland. They are usually found on the face, neck, scalp and back. They are situated in the dermis and vary in size from 0.5 to 5 cm. The cause is unknown.

Seborrhoea

This condition is defined as an excessive secretion of sebum by the sebaceous glands. The glands are enlarged and the skin appears greasy, especially on the nose and the centre zone of the face. The condition may develop into acne vulgaris and is common at puberty, lasting for a few years.

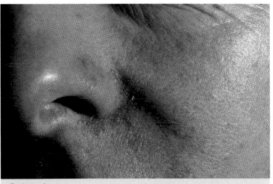

Seborrhoea

Disorders of the sweat glands

Hyperhidrosis

This is the excessive production of sweat, affecting the hands, feet and underarms.

Bacterial infections

Boil

A boil begins as a small inflamed nodule which forms a pocket of bacteria around the base of a hair follicle or a break in the skin. Local injury or lowered constitutional resistance may encourage the development of boils.

Conjunctivitis

This is a bacterial infection following irritation of the conjunctiva of the eye. The inner eyelid and eyeball appear red and sore and there may be a puslike discharge from the eye. The infection spreads by contact with the secretions from the eye of the infected person.

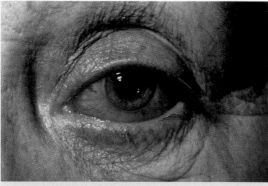
Conjunctivitis

Folliculitis

This bacterial infection occurs in the hair follicles of the skin and appears as a small pustule at the base of a hair follicle. There is redness, swelling and pain around the hair follicle.

Impetigo

This is a superficial, contagious, inflammatory disease caused by streptococcal and staphylococcal bacteria. It is commonly seen on the face and around the ears and its features include weeping blisters which dry to form honey-coloured crusts. This bacteria is easily transmitted by dirty fingernails and towels.

Stye

This is an acute inflammation of a gland at the base of an eyelash, caused by a bacterial infection. The gland becomes hard and tender, and a pus-filled cyst develops at the centre.

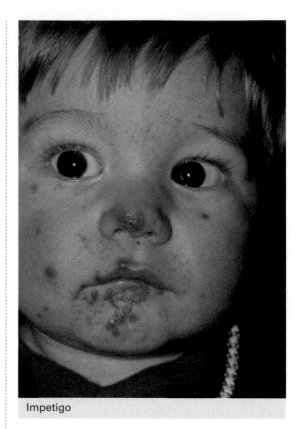
Impetigo

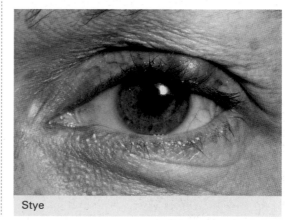
Stye

Viral infections of the skin

Herpes simplex (cold sores)

Herpes simplex is normally found on the face and around the lips. It begins as an itching sensation, followed by erythema and a group of small blisters which then weep and form crusts. This condition will generally persist for approximately two or three weeks, but will reappear at times of stress, ill health or exposure to sunlight.

Herpes simplex (cold sore)

Herpes zoster (shingles)

This is a painful infection along the sensory nerves due to the virus that causes chicken pox. Lesions resemble herpes simplex with erythema and blisters along the lines of the nerves. The areas affected are mostly on the back or upper chest wall. This condition is very painful due to acute inflammation of one or more of the peripheral nerves. Severe pain may persist at the site of shingles for months or even years after the apparent healing of the skin.

Warts

A wart is a benign growth on the skin caused by infection with the human papilloma virus. Plane warts are smooth in texture with a flat top and are usually found on the face, forehead, back of the hands and the front of the knees. Plantar warts or verrucae occur on the soles of the feet and are usually the size of a pea.

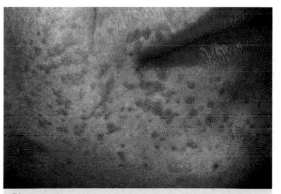

Plane wart

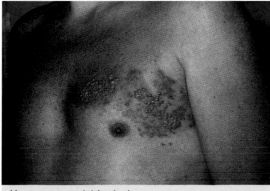

Herpes zoster (shingles)

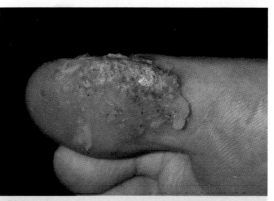

Plantar wart

Fungal infections of the skin

Ringworm

This is a fungal infection of the skin which begins as small red papules that gradually increase in size to form a ring. The affected areas on the body vary in severity from mild scaling to inflamed itchy areas.

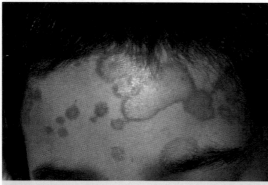

Ringworm

Tinea capitis

This is a type of ringworm and is a fungal infection of the scalp. It appears as painless, round, hairless patches on the scalp. Itching may be present and the lesion may appear red and scaly.

Tinea pedis (athlete's foot)

This is a highly contagious fungal condition which is easily transmitted in damp, moist conditions such as swimming pools, saunas and showers. Athlete's foot appears as flaking skin between the toes which becomes soft and soggy. The skin may also split and the soles of the feet may occasionally be affected.

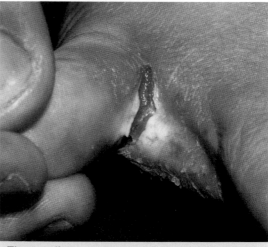

Tinea pedis

Infestation disorders of the skin

Pediculosis (lice)

This condition is commonly known as lice and is a contagious parasitic infection, where the lice live off the blood sucked from the skin. Head lice are frequently seen in young children and, if not dealt with quickly, may lead to a secondary infection of impetigo as a result of scratching. With head lice, nits may be found in the hair. They are pearl-grey or brown, oval structures found on the hair shaft close to the scalp. The scalp may appear red and raw due to scratching.

Body lice are rarely seen. They will occur on an individual with poor personal hygiene and will live and reproduce in seams and fibres of clothing, feeding off the skin. Lesions may appear as papules, scabs and, in severe cases, as pigmented dry, scaly skin. Secondary bacterial infection is often present. A client affected by body lice will complain of itching, especially in the shoulder, back and buttock area.

Scabies

This is a contagious, parasitic skin condition caused by the female mite burrowing into the horny layer of the skin where she lays her eggs. The first noticeable symptom of this condition is severe itching which worsens at night. Papules, pustules and crusted lesions may also develop. Common sites for this infestation are the ulnar borders of the hand, palms of the hands and between the fingers and toes. Other sites include the axillary folds, buttocks, breasts in the female and external genitalia in the male.

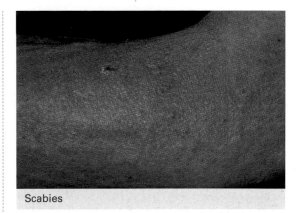
Scabies

Pigmentation disorders

Albinism

This condition is caused by an inherited absence of pigmentation in the skin, hair and eyes, resulting in white hair, and pink skin and eyes. The pink colour is produced by underlying blood vessels which are normally masked by pigment. Other clinical signs of this condition include poor eyesight and sensitivity to light.

Chloasma

This is a pigmentation disorder with irregular areas of increased pigmentation, usually on the face. It commonly occurs during pregnancy and sometimes when taking the contraceptive pill due to stimulation of melanin by the female hormone oestrogen.

Ephelides

Another name for this is freckles. These are small, harmless, pigmented areas of skin. They appear where there is excessive production of the pigment melanin (after exposure to sunlight).

Hypertrophic disorders

Hypertrophic skin disorders refer to conditions which have resulted in an increase of size of a tissue or organ. This is caused by an enlargement of the cells.

Lentlgo

These are also known as liver spots. They are flat, dark patches of pigmentation found mainly in the elderly on skin exposed to light.

Malignant melanoma

A malignant melanoma is a deeply pigmented mole which is life-threatening if it is not recognised and treated promptly. Its main characteristic is a blue-black module which increases in size, shape and colour, and is most commonly found on the head,

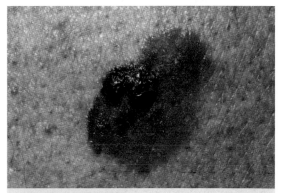
Malignant melanoma

neck and trunk. Overexposure to strong sunlight is a major cause and its incidence is increased in young people with fair skins.

Naevus

A naevus is a birthmark or clearly defined malformation of the skin. There are many different types of naevi:

- **Port wine stain** – also known as a deep capillary naevus. Present at birth and may vary in colour from pale pink to deep purple. It has an irregular shape but is not raised above the skin's surface. Usually found on the face but may also appear on other areas of the body.

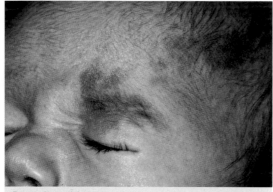
Port wine stain

- **Spider naevi** – a collection of dilated capillaries radiating from a central papule. Often appear during pregnancy or after 'picking a spot'.

- **Strawberry naevus** – usually develops before or shortly after a baby is born, but disappears spontaneously before the child reaches the age of ten. It is a red, raised lump above the skin's surface.

Rodent ulcer

This is a malignant tumour which starts off as a slow-growing, pearly nodule, often at the site of a previous skin injury. As the nodule enlarges, the centre ulcerates and refuses to heal. The centre becomes depressed, and the rolled edges become translucent, revealing many tiny blood vessels. Rodent ulcers do not disappear and if left untreated may invade the underlying bone. This is the most common form of skin cancer.

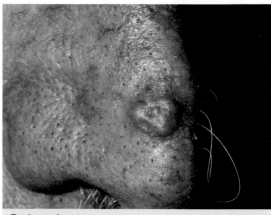
Rodent ulcer

Squamous cell carcinoma

This is a malignant tumour which arises from the prickle-cell layer of the epidermis. It is hard and warty and eventually develops a 'heaped-up, cauliflower' appearance. It is most frequently seen in elderly people.

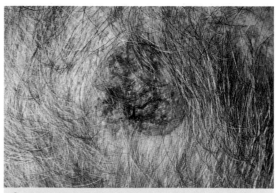
Squamous cell carcinoma

Vitiligo

This condition is present on areas of the skin which lack pigmentation due to the basal cell layer of the epidermis no longer producing melanin. Its cause is unknown.

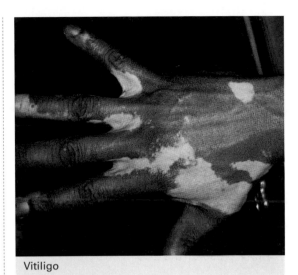

Vitiligo

Inflammatory skin conditions

Contact dermatitis

Dermatitis literally means inflammation of the skin. Contact dermatitis is caused by a primary irritant which causes the skin to become red, dry and inflamed. Substances which are likely to cause this reaction include acids, alkalis, solvents, perfumes, lanolin, detergent and nickels. There may be skin infection as well.

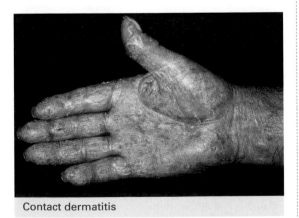

Contact dermatitis

Eczema

This is a mild to chronic inflammatory skin condition characterised by itchiness, redness and the presence of small blisters that may be dry or weep if the surface is scratched. Eczema is non contagious and cause may be genetic or due to internal and external influences. It can cause scaly and thickened skin, mainly at flexures such as the cubital area of the elbows and the back of the knees.

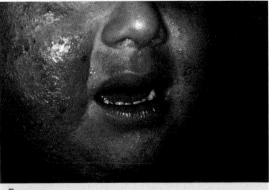

Eczema

Psoriasis

This is a chronic inflammatory skin condition. Psoriasis may be recognised as the

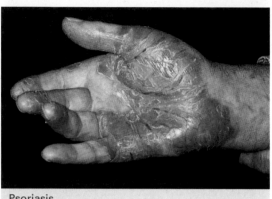

Psoriasis

development of well-defined red plaques, varying in size and shape, and covered by white or silvery scales. Any area of the body may be affected by psoriasis, but the most commonly affected sites are the face, elbows, knees, nails, chest and abdomen. It can also affect the scalp, joints and nails. Psoriasis is aggravated by stress and trauma but is improved by exposure to sunlight.

Seborrhoeic dermatitis

This is a mild to chronic inflammatory disease of hairy areas well supplied with sebaceous glands. Common sites are the scalp, face, axillae and in the groin. The skin may appear to have a grey tinge or a dirty yellow colour. Clinical signs include slight redness, scaling and dandruff in the eyebrows.

Interrelationships with other systems

The skin links to the following body systems:

Skeletal

Vitamin D is produced by the skin on exposure to ultraviolet light. This is needed in bone formation and bone maintenance.

Muscular

Muscles provide a supportive function to the skin. Muscles lie directly under the skin and therefore contribute to skin's tone and elasticity.

Circulatory

Blood clots at the site of an injury and forms a scab on the surface of the skin. This allows the skin to heal and protects underlying structures from any further damage.

Respiratory

Oxygen absorbed into the lungs upon inhalation is delivered to the cells of the skin, hair and nails to aid their renewal.

Nervous

There are numerous sensory nerve endings in the skin that respond to touch, temperature, pain and pressure.

Endocrine

The melanocyte stimulating hormone (MSH) secreted by the central lobe of the pituitary stimulates the production of melanin in the basal cell layer of the skin. The sex hormones (gonadotrophic hormones) influence skin and hair growth during puberty, pregnancy and the menopause.

Digestive

Excess caloric consumption in the daily diet can result in adipose (fatty tissue) being stored in the subcutaneous layer of the skin.

Urinary

Water is lost from the skin as sweat. The kidneys regulate fluid balance in the body to prevent the skin from becoming dehydrated.

Key words associated with the skin

epidermis	reticular layer	medulla
dermis	adipose tissue	lanugo hair
subcutaneous layer	collagen	vellus hair
basal layer (stratum germinativum)	elastin	terminal hair
	reticular fibres	anagen
prickle-cell layer (stratum spinosum)	hair	catagen
	sebaceous gland	telogen
granular layer (stratum granulosum)	eccrine gland	nail plate
	apocrine gland	nail bed
clear layer (stratum lucidum)	erector pili muscle	matrix
horny layer (stratum corneum)	hair shaft	nail wall
	hair root	nail groove
cell regeneration	hair bulb	cuticle
keratinisation	cuticle	lunula
desquamation	cortex	free edge
papillary layer		

Summary of the skin

- The **skin** and the appendages derived from it (**hair, glands** and **nails**) make up the skin.

- The skin is one of the larger organs in the body.

- Functions of the skin include **protection, regulation of body temperature, sensation, excretion, storage, absorption** and **vitamin D production**.

- The principal parts of the skin are the outer **epidermis** and the inner **dermis**. Beneath the dermis lies the **subcutaneous layer**.

- The epidermis is the most superficial part and consists of five layers, from deepest to superficial: **basal cell layer** (stratum germinativum), **prickle-cell layer** (stratum spinosum), **granular layer** (stratum granulosum), **clear layer** (stratum lucidum) and **horny layer** (stratum corneum).

- **Cell regeneration** occurs continuously in the **basal cell layer** and produces all other layers.

- It takes approximately a month for a new cell to complete its journey from the **basal cell layer**, where it is reproduced, to the **granular layer**, where it is keratinised, to the **horny layer**, where it is **desquamated** or shed.

- The **dermis** is the deeper layer of the skin and provides support, strength and elasticity.

- It has a superficial **papillary** layer and a deeper **reticular** layer.

- The superficial **papillary** layer consists of **adipose connective tissue**, **dermal papillae**, **nerve endings** and a network of **blood** and **lymphatic capillaries**.

- The deeper **reticular layer** consists of tough **fibrous connective tissue** and contains **collagen**, **elastin** and **reticular** fibres.

- Appendages of the skin include the **hair**, glands (**sebaceous and sweat**) and **nails**.

- The **hair** is a dead, keratinised structure and is divided into three parts: **hair shaft**, **root** and **bulb**.

- The role of a hair is protection.

- Internally the hair has three layers, from the outer to inner layer: **cuticle**, **cortex** and **medulla**.

- The **matrix** and **hair bulb** is the area of mitotic activity for the hair cells.

- There are three main types of hair in the body: **lanugo**, **vellus** and **terminal**.

- Each hair has its own hair growth cycle.

- **Anagen** is the active growing stage, **catagen** is the transitional stage from active to resting and **telogen** is the short resting stage.

- Nails are made up mainly of **keratin** and are a modification of the horny and clear layers of the epidermis.

- The two main functions of the nail are protection for the fingers and toes, and for manipulation of objects.

- Parts of the nail's anatomical structure include the **nail plate**, **nail bed**, **matrix**, **nail wall**, **nail groove**, **cuticle**, **lunula** and **free edge**.

- Nail growth occurs from the nail **matrix** by cell division.

- As new cells are produced in the matrix, older cells are pushed forward and are hardened by **keratinisation** to form the hardened **nail plate**.

- Other related structures of the skin are the **erector pili muscle** and the glands.

- The **erector pili muscle** is the weak muscle associated with hair and will contract when you are cold or experiencing emotions such as fright or anxiety.

- **Sebaceous glands** are also known as oil glands. They have ducts and are attached to hair follicles.

- They secrete **sebum** which is mildly antibacterial and antifungal to lubricate the hair and the epidermis.

- **Sweat glands** are located in the **dermis** and secrete **sweat**.

- There are two types of sweat glands, **eccrine and apocrine**.

- **Eccrine glands** are the most numerous and are found in largest concentration in the palms of the hands and soles of the feet.

- **Apocrine glands** are attached to the hair follicles and are located in the axillae and groin.

- Factors affecting the skin include diet, water intake, sleep, stress and tension, exercise, alcohol, smoking, medication, chemicals, climate, environment, hormones and age.

Multiple-choice questions

The skin

1. **In which of the following layers are epidermal cells constantly being reproduced?**
   a horny layer
   b granular layer
   c basal cell layer
   d clear layer

2. **Desquamation occurs in which layer of the epidermis?**
   a clear layer
   b basal cell layer
   c prickle-cell layer
   d horny layer

3. **The correct order of the layers of the epidermis from deepest to most superficial is as follows:**
   a basal cell layer, prickle-cell layer, clear layer, granular layer, horny layer
   b basal cell layer, prickle-cell layer, granular layer, clear layer, horny layer
   c basal cell layer, horny layer, clear layer, granular layer, prickle-cell layer
   d basal cell layer, granular layer, prickle-cell layer, clear layer, horny layer

4. **The erector pili muscle affects:**
   a the hair
   b the sebaceous gland
   c motor nerves
   d capillary network

5. **The function of the sebaceous gland is to:**
   a secrete sweat
   b secrete sebum
   c regulate temperature
   d insulate

6. **Which of the following is responsible for making the skin pigmentation darker?**
   a keratin
   b melanin
   c sebum
   d carotene

7. **Which of the following statements is true?**
   a the epidermis does not have its own system of blood vessels
   b the muscle layer in the skin lies beneath the dermis
   c sweat glands open into a hair follicle
   d melanin is produced by special cells called phagocytes

8. **Which of the following is not a function of the skin?**
   a protection
   b temperature regulation
   c transportation
   d excretion

9. **The tough protein found in the epidermis, hair and nails is:**
   a melanin
   b keratin
   c reticulin
   d none of the above

10. **In which layer of the skin would you find collagen fibres?**
   a the papillary layer
   b the basal cell layer
   c the subcutaneous layer
   d the reticular layer

11. The function of the subcutaneous layer is to:

a support blood vessels
b insulate the body
c support nerve endings
d all of the above

12. Which of the following statements is false?

a when the body is losing heat the blood capillaries near the surface constrict
b the evaporation of sweat from the surface of the skin heats the body
c when heat needs to be retained the erector pili muscle contracts
d adipose tissue in the dermis and subcutaneous layer insulates against heat loss

13. Which of the following is a fungal infection in the skin?

a impetigo
b herpes simplex
c ringworm
d herpes zoster

14. Which of the following would represent a contraindication to therapy treatments due to cross-infection?

a tinea pedis
b conjunctivitis
c scabies
d all of the above

15. What is meant by the term erythema in the skin?

a a mass of dilated capillaries in the skin
b a small, raised elevation in the skin
c a mark left on the skin after a wound has healed
d reddening of the skin due to the dilation of blood capillaries

16. The most lethal type of skin cancer is:

a squamous cell carcinoma
b rodent ulcer
c basal cell carcinoma
d malignant melanoma

The hair

1. Hair growth occurs from the:

a cuticle
b cortex
c medulla
d matrix

2. Hair grows from a saclike depression called the:

a hair shaft
b hair root
c hair follicle
d hair bulb

3. The three layers of the internal hair structure are:

a hair shaft, hair root and hair bulb
b cuticle, cortex and medulla
c outer root sheath, inner root sheath and hair shaft
d none of the above

4. Where are terminal hairs found on the body?

a on the scalp
b under the arms
c pubic region
d all of the above

5. Which of the following provides a crucial source of nourishment for hair?

a connective tissue sheath
b dermal papilla
c outer root sheath
d inner root sheath

6. **Which of the following statements is true in relation to the hair growth cycle?**

a catagen is a short resting stage

b anagen lasts approximately two to four weeks

c anagen is the active growing stage

d in anagen the hair separates from the dermal papilla and moves slowly up the follicle

The nail

1. **The substance of the nail is mainly composed of:**

a calcium

b keratin

c blood

d none of the above

2. **The area of the nail where living cells are produced is:**

a nail bed

b matrix

c lunula

d none of the above

3. **The part of the nail that protects the matrix and provides a protective seal against bacteria is:**

a cuticle

b nail wall

c nail groove

d lunula

4. **Which of the following statements is true?**

a the average growth rate of a nail is 6 mm per month

b fingernails have a slower growth rate than toenails

c nail growth occurs from the matrix by cell division

d it takes approximately three months for cells to travel from the lunula to the free edge of the nail

5. **All nutrients are supplied to the nail via which layer of the skin?**

a the epidermis

b the subcutaneous layer

c the dermis

d none of the above

6. **A common nail disease characterised by inflammation and bacterial infection of the skin surrounding the nail is:**

a pterygium

b onychomycosis

c paronychia

d leuconychia

the skeletal system

Introduction

The skeleton is made up of no fewer than 206 individual bones which collectively form a strong framework for the body. Bones are like landmarks in the body and by tracing their outlines you can be accurate in describing the position of muscles, glands and organs in the body. Learning the positions of the bones of the skeleton is essential for learning the position of the muscles, as bones must have muscle attachments to enable them to move.

Objectives

By the end of this chapter you will be able to recall and understand the following knowledge:

- functions of the skeleton
- structure of bone
- growth and development of bone
- different types of bone in the body
- names and positions of the bones of the skeleton
- different types of joints and their range of movement
- the importance of good posture
- postural deformities
- the interrelationships between the skeletal and other body systems
- disorders of the skeletal system.

The skeleton comprises bones that provide support and protection for our body. Bones must, however, be linked together in order to facilitate their supportive role and to allow movement. Joints provide the link between bones of the skeletal system. Joints, therefore, serve a dual purpose. They hold bones together via ligaments, offering stability for the joint and giving the skeletal system more flexibility by facilitating movement which is assisted by associated muscles and tendons.

Functions of the skeleton

The skeletal system is made up of all types of bones which form the skeleton or bony framework of the body. Before learning the individual bones of the skeleton, it is important to understand the functions of the skeleton as a whole.

Support

The skeleton bears the weight of all other tissues. Without it we would be unable to stand up. Consider the bones of the vertebral column, pelvis, feet and legs, which all support the weight of the body.

Shape

The bones of the skeleton give shape to structures such as the skull, thorax and limbs.

Protection of vital organs and delicate tissue

The skeleton surrounds vital organs and tissue with a tough and resilient covering, such as the ribcage protecting the heart and lungs and the vertebral column protecting the spinal cord.

Attachments for muscles and tendons

Bones are like anchors which allow the muscle to function efficiently.

Movement

This happens as a result of the coordinated action of muscles on bones and joints. Bones are, therefore, levers for muscles.

Formation of blood cells

These develop in red bone marrow found in cancellous bone tissue.

Mineral reservoir

The skeleton acts as a storage depot for important minerals, such as calcium, which can be released when needed for essential metabolic processes like muscle contraction and the conduction of nerve impulses.

The structure of bone

Bone is one of the hardest types of connective tissue in the body and when fully developed is composed of water, calcium salts and organic matter. Bone tissue is a type of living tissue that is made from special cells called osteoblasts. There are two main types of bone tissue: **compact** and **cancellous**. All bones have both types of tissue, the amount being dependent on the type of bone.

Compact (dense) bone

This is the hard portion of the bone that makes up the main shaft of the long bones and the outer layer of other bones. It protects spongy bone and provides a firm framework for the bone and body. The bone cells in this type of bone (osteocytes) are located in concentric rings around a central haversian canal, through which nerves, blood and lymphatic vessels pass.

Cancellous (spongy) bone

In contrast, this is lighter in weight than compact bone. It has an open, spongelike appearance and is found at the ends of long bones or at the centre of other bones. It does not have a haversian system but consists of a weblike arrangement of spaces that are filled with red bone marrow and separated by the thin processes of bone. Blood vessels run through every layer of cancellous bone, conveying nutrients and oxygen.

Bone marrow

Bones contain two types of marrow – red and yellow:

- Red marrow manufactures red blood cells. It is found at the end of long bones and at the centre of other bones of the thorax and pelvis.
- Yellow marrow is found chiefly in the central cavities of long bones.

Except for the ends that form joints, bones are covered with a thin membrane of connective tissue called the periosteum. The outer layer of the periosteum is extremely dense and contains a large number of blood vessels. The inner layer contains osteoblasts and fewer blood vessels. The periosteum provides attachment for muscles, tendons and ligaments.

The development of bone

The process of bone development is called **ossification**. The bones of a foetus are made of cartilage rods that are changed into bone due to ossification as the child develops and grows. This process begins in the embryo near the end of the second month and is not complete until about the twenty-fifth year of life.

Ossification takes place in three stages:

1 The cartilage-forming cells, called chondrocytes, enlarge and arrange themselves in rows similar to the bone they will eventually form.

2 Calcium salts are then laid down by special bone-building cells called osteoblasts.

3 A second set of cells called osteoclasts, known as cartilage-destroying cells, brings about an antagonistic action, enabling the absorption of any unwanted bone.

A fine balance of osteoblast and osteoclast activity helps to maintain the formation of normal bone. Osteocytes are mature bone cells that maintain bone during our lifetime.

Cartilage

Cartilage is a dense, connective tissue that consists of collagen and elastin fibres embedded in a strong, gel-like substance. It is a flexible and durable type of tissue, cushioning and absorbing shock, thereby preventing direct transmission to bones.

There are three types of cartilage:

- **hyaline** – covers the articular bone surfaces
- **fibrous** – a strong and rigid type of cartilage between the discs of the spine
- **elastic** – a very flexible type of cartilage found in the auditory canal of the ear.

Cartilage has no blood supply and, therefore, does not repair or renew itself as easily as bone.

Ligaments

Ligaments are dense, strong, flexible bands of white fibrous connective tissue that link bones together at a joint. They are inelastic but flexible, strengthening the joint and allowing the bones to move freely within a safe range of movement.

Tendons

Tendons are tough, white fibrous cords of connective tissue that attach muscles to the periosteum (fibrous covering) of a bone. Tendons enable bones to move when skeletal muscles contract.

Types of bone

Bones are classified according to their shape. They are classified as **long bones**, **short bones**, **flat bones**, **irregular bones** and **sesamoid bones**.

Long bones

Long bones have a long shaft (diaphysis) and one or more endings or swellings (epiphysis). Smooth hyaline cartilage covers the articular surfaces of the shaft endings. Between the diaphysis and epiphysis of growing bone is a flat plate of hyaline cartilage called the epiphyseal cartilage or growth plate. This is the site of bone growth and as fast as this cartilage grows it is turned into bone, allowing the bone to continue to grow in length.

A growth spurt is often seen during puberty through the influence of the sex hormones oestrogen and testosterone, both of which promote the growth of long bone. At around 25 years of age the entire plate becomes ossified.

All bones of the limbs are long bones (except the wrist and ankle bones).

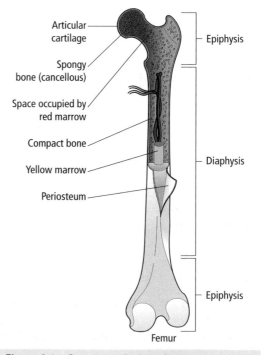

Figure 3.1   Structure of a long bone

Key note

Children's bones are more flexible as their bodies contain more cartilage and soft bone cells, since complete calcification has not yet taken place. In older adults this process is reversed, as bone cells outnumber cartilage cells and bone becomes more brittle as it contains more minerals and fewer blood vessels. This explains why elderly people's bones are more prone to fracture and slower to heal.

Short bones

These bones are generally cube-shaped, with their lengths and widths being roughly equal. The bones of the wrist and the ankle are examples of short bones.

Flat bones

Flat bones are platelike structures with broad surfaces. Examples include the ribs and the scapulae.

Irregular bones

Irregular bones have a variety of shapes. Examples include the vertebrae and some of the facial bones.

Sesamoid bones

These are small, rounded bones embedded in a tendon. The largest sesamoid bone is the patella, which is embedded in the quadriceps femoris tendon.

The bones of the skeleton

The skeletal system is divided into two parts: the axial skeleton made up of 80 bones and the appendicular skeleton made up of 126 bones.

The axial skeleton forms the main axis or central core of the body and consists of the following parts:

■ skull

■ vertebral column

■ sternum

■ ribs.

The appendicular skeleton supports the appendages or limbs and gives them attachment to the rest of the body. It consists of the following parts:

■ shoulder girdle

■ bones of the upper limbs

■ bones of the lower limbs

■ bones of the pelvic girdle.

The skull

The skull rests on the upper end of the vertebral column, weighs around 11 pounds and consists of 22 bones. Eight bones make up the skull or cranium and 14 bones form the facial skeleton. The skull encloses and protects the brain and provides a surface attachment for various muscles of the skull.

The eight bones of the skull are as follows:

■ One **frontal** bone forms the anterior part of the roof of the skull, the forehead and the upper part of the orbits or eye sockets.

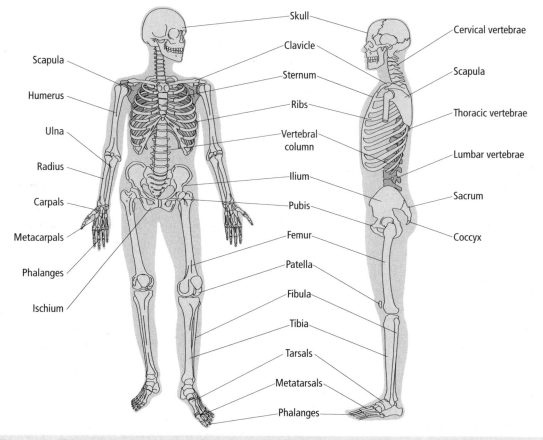

Figure 3.2   Bones of the skeleton (anterior and side)

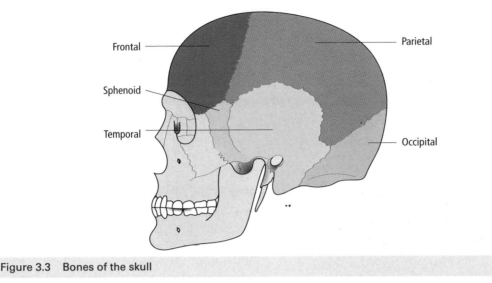

Figure 3.3 Bones of the skull

Within the frontal bones are the two frontal sinuses, one above each eye, near the midline.

■ Two **parietal** bones form the upper sides of the skull and the back of the roof of the skull.

■ Two **temporal** bones form the sides of the skull below the parietal bones and above and around the ears. The temporal bone contributes to part of the cheekbone via the zygomatic arch, formed by the zygomatic and temporal bone. Located behind the ear and below the line of the temporal bones are the mastoid processes to which the stermomastoid muscles of the neck are attached.

■ One **sphenoid** bone is located in front of the temporal bone and serves as a bridge between the cranium and the facial bones. It articulates with the frontal, temporal, occipital and ethmoid bones.

■ One **ethmoid** bone forms part of the wall of the orbit, the roof of the nasal cavity and part of the nasal septum.

■ The **occipital** bone forms the back of the skull.

Key note

There are many openings present in the bones of the skull which act as passages for blood vessels and nerves entering and leaving the cranial cavity. An example is the large opening at the base of the skull called the foramen magnum through which the spinal cord and blood vessels pass to and from the brain.

The bones of the face

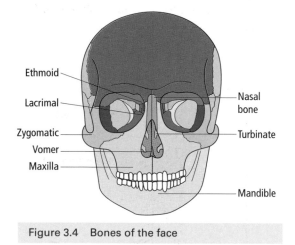

Figure 3.4 Bones of the face

There are 14 facial bones in total. These are mainly in pairs, one on either side of the face:

- **Two maxillae** – these are the largest bones of the face and they form the upper jaw and support the upper teeth. The two maxillary bones fuse to become one early in life. An important part of the maxillae are the maxillary sinuses which open into the nasal cavity.

- **One mandible** – this is the only moveable bone of the skull and forms the lower jaw and supports the lower teeth. The mandible is the largest and heaviest bone in the skull.

- **Two zygomatic** – these are the most prominent of the facial bones and form the cheekbones.

- **Two nasal** – these small bones form the bridge of the nose.

- **Two lacrimal** – these are the smallest of the facial bones and are located close to the medial part of the orbital cavity.

- **Two turbinate** – these are layers of bone located either side of the outer walls of the nasal cavities.

- **One vomer** – this is a single bone at the back of the nasal septum.

- **Two palatine** – these are L-shaped bones which form the anterior part of the roof of the mouth.

The sinuses

There are four pairs of air-containing spaces in the skull and face called the sinuses. The function of the sinuses is to lighten the head, provide mucus and act as a resonance chamber for sound. The pairs of sinuses are named according to the facial bones by which they are located. They are the **frontal** sinuses, the **sphenoidal** sinuses, the **ethmoidal** sinuses and the **maxillary** sinuses, which are the largest.

The vertebral column

The vertebral column lies on the posterior of the skeleton, extending from the skull to the pelvis, providing a central axis to the body. It consists of 33 individual irregular bones called vertebrae. However, the bones of the base of the vertebral column, the sacrum and coccyx, are fused to give 24 movable bones in all.

The vertebral column is made up of the following:

- **Seven cervical vertebrae** – these are the vertebrae of the neck. The top two vertebrae, C1 the atlas and C2 the axis, are part of a pivot joint which allows the head and neck to move freely. These are the smallest vertebrae in the vertebral column.

■ **Twelve thoracic vertebrae** – these are the vertebrae of the mid-spine and lie in the thorax where they articulate with the ribs. These vertebrae lie flatter and downwards to allow for muscular attachment of the large muscle groups of the back. They can be felt easily as you run your fingers down the spine.

■ **Five lumbar vertebrae** – these lie in the lower back and are much larger in size than the vertebrae above them as they are designed to support more body weight. These vertebrae can be easily felt on the lower back due to their large shape and width.

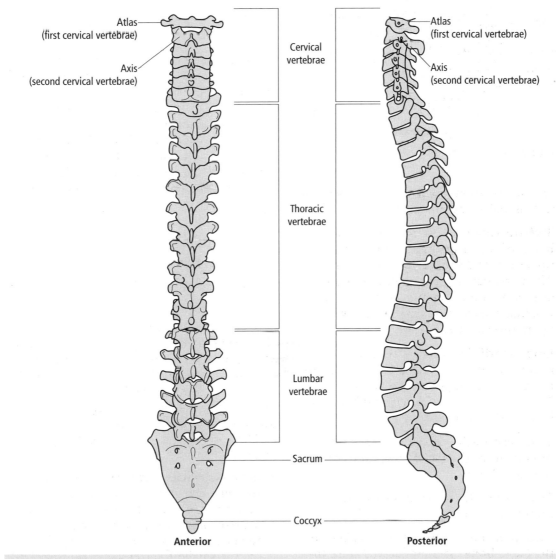

Atlas
(first cervical vertebrae)

Axis
(second cervical vertebrae)

Cervical
vertebrae

Thoracic
vertebrae

Lumbar
vertebrae

Atlas
(first cervical vertebrae)

Axis
(second cervical vertebrae)

Sacrum

Coccyx

Anterior

Posterior

Figure 3.5   Bones of the vertebral column

- **Five sacral vertebrae** – the sacrum is a very flat triangular-shaped bone and lies in between the pelvic bones. It is made up of five bones which are fused together. A characteristic feature of the sacrum are the eight sacral holes. It is through these holes that nerves and blood vessels penetrate.
- **Four coccygeal vertebrae** – these are made up of four bones which are fused together and are sometimes referred to as the tailbone.

Key note

In between the vertebrae lies a padding of fibrocartilage called the intervertebral discs. These give the vertebrae a certain degree of flexibility and also act as shock absorbers in between the vertebrae, cushioning any mechanical stress that may be placed on them.

Functions of the vertebral column

Now we have covered the individual structure of the vertebrae, let us consider the functions of the vertebral column as a whole:

- The vertebral column provides a strong and slightly flexible axis to the skeleton.
- By way of its different-shaped vertebrae, with their roughened surfaces, it is able to provide a surface for the attachment of muscle groups.
- The vertebral column also has a protective function as it protects the delicate nerve pathways of the spinal cord.

Therefore, the vertebral column mirrors the primary functions of the skeleton in its supportive and protective roles.

The thoracic cavity

This is the area of the body enclosed by the ribs, providing protection for the heart and lungs.

Essential parts contained within this cavity include:

- the sternum
- the ribs
- 12 thoracic vertebrae.

The sternum

This is commonly referred to as the breast bone and is a flat bone lying just beneath the skin in the centre of the chest. The sternum is divided into three parts:

- **manubrium**, the top section
- **main body**, the middle section
- **xiphoid process**, the bottom section.

The top section of the sternum articulates with the clavicle and the first rib. The middle section articulates with the costal cartilages that link the ribs to the sternum. The bottom section provides a point of attachment for the muscles of the diaphragm and the abdominal wall.

The ribs

There are 12 pairs of ribs. They articulate posteriorly with the thoracic vertebrae. Anteriorly, the first 10 pairs attach to the sternum via the costal cartilages, the first 7 directly (known as the true ribs), the remaining 3 indirectly (known as the false ribs). The last 2 ribs have no anterior attachment and are called the floating ribs.

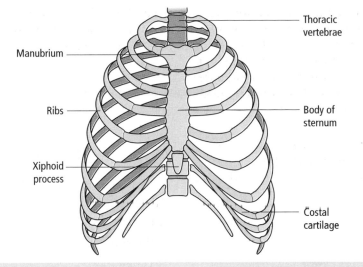

Figure 3.6   The thoracic cavity

The appendicular skeleton

This consists of the shoulder girdle, the bones of the upper and lower limbs and the pelvic girdle.

The shoulder girdle

The shoulder girdle connects the upper limbs with the thorax and consists of four bones – two **scapula** and two **clavicle**.

The **scapula** is a large, flat bone, triangular in outline, which forms the posterior part of the shoulder girdle. It is located between the second and the seventh ribs. The scapula articulates with the clavicle and the humerus and serves as a point of muscle attachment which connects the shoulder girdle with the trunk and upper limbs.

The **clavicle** is a long, slender bone with a double curve. It forms the anterior portion of the shoulder girdle. At its medial end it

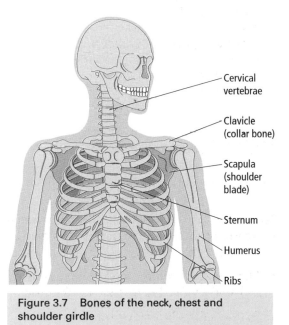

Figure 3.7   Bones of the neck, chest and shoulder girdle

articulates with the top part of the sternum and at its lateral end it articulates with the scapula. The clavicle acts as a brace to hold the arm away from the top of the thorax.

The clavicle provides the only bony link between the shoulder girdle and the axial skeleton. The arrangement of bones and the muscle attached to the scapula and the clavicle allow for a considerable amount of movement of the shoulder and the upper limbs.

The upper limb

The upper limb consists of the following bones:

- humerus
- radius
- ulna
- carpals
- metacarpals
- phalanges.

Humerus

The humerus is the long bone of the upper arm. The head of the humerus articulates with the scapula, forming the shoulder joint. The distal end of the bone articulates with the radius and ulna to form the elbow joint.

Radius and ulna

The ulna and radius are the long bones of the forearm. The two bones are bound together by a fibrous ring which allows a rotating movement in which the bones pass over each other. The ulna is the bone of the little finger side and is the longer of the two forearm bones. The radius is situated on the thumb side of the forearm and is shorter than the ulna. The joint between the ulna and the radius permits a movement called pronation. This is when the radius moves obliquely across the ulna so that the thumb side of the hand is closest to the body. The movement called supination takes the thumb side of the hand to the lateral side.

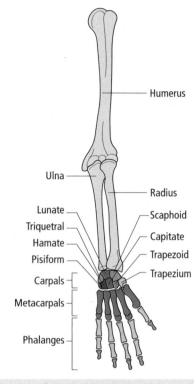

Figure 3.8　Bones of the upper limb

The radius and the ulna articulate with the humerus at the elbow and the carpal bones at the wrist.

The wrist and hand

Carpals

The wrist consists of eight small bones of irregular size which are collectively called carpals. They fit closely together and are held in place by ligaments. The carpals are arranged in two groups of four. Those of the upper row articulate with the ulna and the radius and the lower row articulates with the metacarpals. The upper row nearest the forearm consists of scaphoid, lunate, triquetral and pisiform. The lower row is made up of trapezium, trapezoid, capitate and hamate.

Metacarpals

There are five long metacarpal bones in the palm of the hand. Their proximal ends articulate with the wrist bones and the distal ends articulate with the finger bones.

Phalanges

There are 14 phalanges. These are the finger bones; two are in the thumb or pollex and three are in each of the other digits.

The lower limb

The lower limb consists of the following bones:

- femur
- patella
- tibia
- fibula
- tarsals
- metatarsals
- phalanges.

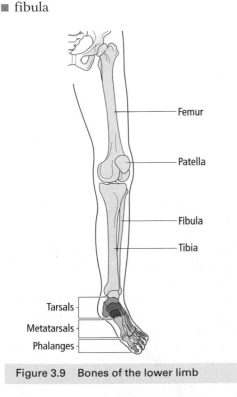

Femur

Patella

Fibula

Tibia

Tarsals

Metatarsals

Phalanges

Figure 3.9   Bones of the lower limb

Femur

The femur is the bone of the thigh. It is the longest bone in the body and has a shaft and two swellings at each end. The proximal swelling has a rounded head like a ball which fits into the socket of the pelvis to form the hip joint. Below the neck are swellings called trochanters which are sites for muscle attachment. The distal ends of the femur articulate with the patella or kneecap.

The patella

The patella, or kneecap, is located anterior to the knee joint. Its main function is to provide stabilisation, cushion the hinge joint at the knee and protect the knee by shielding it from impact.

Tibia and fibula

The tibia and fibula are long bones of the lower leg. The tibia is situated on the anterior and medial side of the lower leg. It has a large head where it joins the knee joint, and the shaft leads down to form part of the ankle. The tibia is the larger of the two bones of the lower leg and thus carries the weight of the body. The fibula is situated on the lateral side of the tibia in the lower leg and is the shorter and thinner of the two bones. The end of the fibula forms part of the ankle on the lateral side.

The foot

Tarsals

There are seven bones in the foot which are collectively called the tarsals. Each tarsal is an irregular bone that slides minutely over the next bone to collectively provide motion. The individual tarsals are as follows:

- **Talus** – the talus bone is the main tarsal. It articulates with the tibia and fibula to form the ankle joint. The talus is significant in that it bears the weight of the entire body when standing or walking.

- **Calcaneum** – the calcaneum is also known as the heel bone. It is the largest and most posterior tarsal bone. The calcaneum is an important site for attachment of muscles of the calf.

- **Cuboid** – the cuboid is situated between the fourth and fifth metatarsals and the calcaneum on the lateral (outer) border of the foot.

- **Cuneiform** – there are three cuneiform bones which are located between the navicular bone and the first three metatarsal bones. They are numbered medially to laterally from I through to III.

- **Navicular** – the navicular bone is situated between the talus bone and the three cuneiforms.

- **Metatarsals** – there are five metatarsals forming the dorsal surface of the foot.

- **Phalanges** – 14 phalanges form the toes, 2 of which are in the hallux or big toe and 3 to each of the other digits.

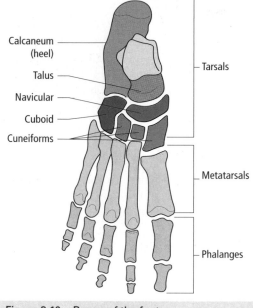

Figure 3.10   Bones of the foot

Arches of the foot

The bones of the feet form arches which are designed to support body weight and provide leverage when walking. The arches of the foot are maintained by ligaments and muscles. They give the foot resilience in bearing the body's weight when running or walking. The arches of the foot are:

- **Medial longitudinal arch** – this runs along the medial side of the foot from the calcaneum bone to the end of the metatarsals.

- **Lateral longitudinal arch** – this runs along the lateral side of the foot from the calcaneum bone to the end of the metatarsals.

- **Transverse arch** – this runs between the medial and lateral aspect of the foot and is formed by the navicular, three cuneiforms and the bases of the five metatarsals.

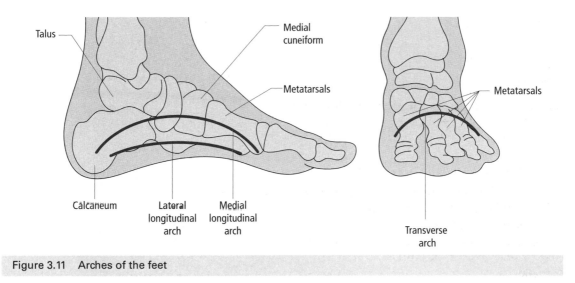

Figure 3.11   Arches of the feet

The pelvic girdle

The pelvic girdle consists of two hip bones which are joined together at the back by the sacrum and at the front by the symphysis pubis.

Each hip bone consists of three separate bones which are fused together. They are the:

 ilium   ■ ischium

■ pubis.

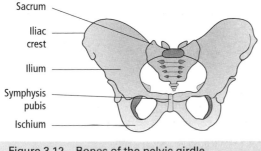

Figure 3.12   Bones of the pelvic girdle

Ilium

The ilium is the largest and most superior pelvic bone in the pelvic girdle. Its upper border is called the iliac crest which is an important site of attachment for muscles of the anterior and posterior abdominal walls.

Ischium

The ischium is the bone at the lower end and forms the posterior part of the pelvic girdle.

Key note

The ischial tuberosity is a bony protrusion which is the part of the ischium that you sit on. It receives the weight of the body when sitting and provides muscle attachments for the muscles such as the hamstrings and the adductors. Poor sitting posture can result from sitting on your sacrum instead of the ischial tuberosity.

Pubis

The pubis is the collective name for the two pubic bones in the most anterior portion of the pelvis. The two pubic bones resemble a wishbone and are linked via a piece of cartilage called the symphysis pubis. The pubic bones provide attachment sites for some of the abdominal muscles and fasciae.

Functions of the pelvic girdle

Like the vertebral column, the pelvic girdle mirrors the primary functions of the skeleton – it has a role in supporting the vertebral column and the body's weight and offers protection by encasing delicate organs such as the uterus and bladder.

Joints

A joint is formed where two or more bones or cartilage meet and is otherwise known as an articulation. Where a bone is a lever in a movement, the joint is the fulcrum or the support which steadies the movement and allows the bone to move in certain directions.

Types of joint

Joints are classified according to the degree of movement possible at each one. There are three main joint classifications:

- **fibrous** – no movement is possible (also known as fixed joints)
- **cartilaginous** – slight movement is possible
- **synovial** – freely movable joints.

Fibrous joints

These are immovable joints with tough fibrous tissue between the bones. Often the edges of the bones are dovetailed, as in the sutures of the skull. Some examples of fibrous joints include the joints between the teeth and between the maxilla and mandible of the jaw.

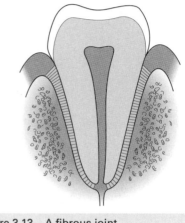

Figure 3.13   A fibrous joint

Cartilaginous joints

These are slightly movable joints which have a pad of fibrocartilage between the ends of the bones making the joint. The pad acts as a shock absorber. Some examples of cartilaginous joints are those between the vertebrae of the spine and at the symphysis pubis, in between the pubis bones.

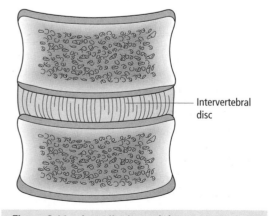

Intervertebral disc

Figure 3.14   A cartilaginous joint

Synovial joints

These are freely movable joints which have a more complex structure than the fibrous or cartilaginous joints. Before looking at the different types of synovial joints it is important to have an understanding of the general structure of a synovial joint.

The general structure of a synovial joint

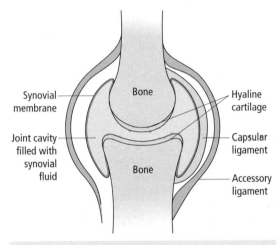

Synovial membrane — Bone — Hyaline cartilage — Capsular ligament — Joint cavity filled with synovial fluid — Bone — Accessory ligament

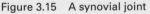

Figure 3.15   A synovial joint

- A synovial joint has a space between the articulating bones which is known as the synovial cavity.

- The surface of the articulating bones is covered by hyaline cartilage which is supportive to the joint by providing a hard-wearing surface for the bones to move against one another with the minimum of friction.

- The synovial cavity and the cartilage are encased within a fibrous capsule which helps to hold the bones together to enclose the joint. This joint capsule is reinforced by tough sheets of connective tissue called ligaments which bind the articular ends of bones together.

- The joint capsule is reinforced enough to allow strength to resist dislocation, but is flexible enough to allow movement at the joint.

- The inner layer of the joint capsule is formed by the synovial membrane which secretes a sticky, oily fluid called synovial fluid which lubricates the joint and nourishes the hyaline cartilage.

- As the hyaline cartilage does not have a direct blood supply, it relies on the synovial fluid to deliver its oxygen and nutrients and to remove waste from the joint which is achieved via the synovial membrane.

Types of synovial joints

Synovial joints are classified into six different types according to their shape and the movements possible at each one. The degree of movement possible at each synovial joint is dependent on the type of synovial joint and its articulations.

Ball-and-socket joint

A ball-and-socket joint allows movement in many directions around a central point. This type of joint is formed when the rounded head of one bone fits into a cup-shaped cavity of another bone. Angular movements possible at this joint include flexion, extension, adduction, abduction, rotation and circumduction. Ball-and-socket joints allow the greatest freedom of movement but are also the most easily dislocated. Examples include the hip and the shoulder joints.

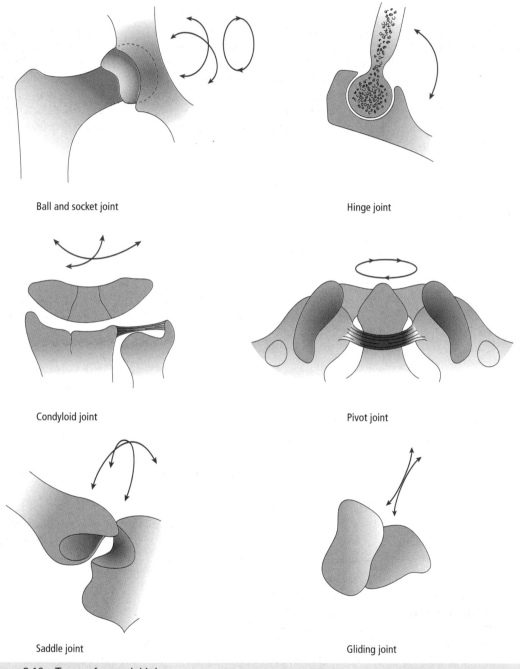

Ball and socket joint

Hinge joint

Condyloid joint

Pivot joint

Saddle joint

Gliding joint

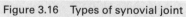

Figure 3.16   Types of synovial joint

Hinge joint

This type of joint is where the rounded surface of one bone fits the hollow surface of another bone. Movement is only possible in one direction. A hinge joint allows flexion and extension, changing the angle of the bones at the joint like a door hinge. Examples include the knee and the elbow joints and the joints in between the phalanges.

Condyloid joint

The joint surfaces of a condyloid joint are shaped so that the concave surface of one bone can slide over the convex surface of another bone in two directions. Although a condyloid joint allows movement in two directions, one movement dominates. Movements possible at a condyloid joint include flexion, extension, adduction and abduction. Examples include the wrist joint and the joint between the metacarpals and phalanges (metacarpophalangeal joints).

Gliding joint

Gliding joints permit gliding between two or more bones. They are often referred to as synovial plane joints as they occur where two flat surfaces of bone slide against one another. Gliding joints allow only a gliding motion in various planes. Examples include the joints between the vertebrae and the sacroiliac joint.

Pivot joint

A pivot joint occurs where a process of bone rotates in a socket. One component is shaped like a ring and the other component is shaped so that it can rotate within the ring. A pivot joint only permits rotation. Examples include the joint between the first and second cervical vertebrae (atlas and axis) and the joint at the proximal ends of the radius and the ulna.

Saddle joint

This type of joint is shaped like a saddle. Its articulating surfaces of bone have both rounded and hollow surfaces so that the surface of one bone fits the complementary surface of the other. Movements possible at this joint include flexion, extension, adduction, abduction and a small degree of axial rotation.

Glossary of angular movements possible at joints

Flexion: bending of a body part at a joint so that the angle between the bones is decreased

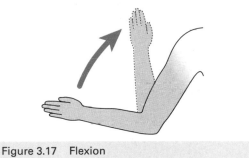

Figure 3.17   Flexion

Extension: straightening of a body part at a joint so that the angle between the bones is increased

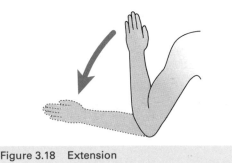

Figure 3.18   Extension

Dorsiflexion: upward movement of the foot so that feet point upwards

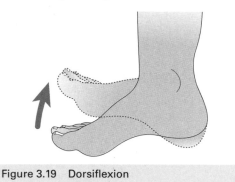

Figure 3.19   Dorsiflexion

Plantar flexion: downward movement of the foot so that feet face downwards towards the ground

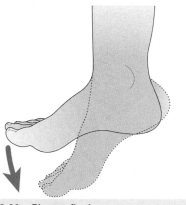

Figure 3.20   Plantar flexion

Adduction: movement of a limb towards the midline

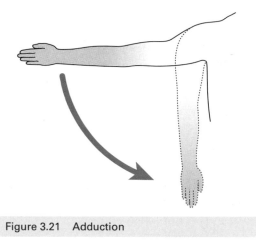

Figure 3.21   Adduction

Abduction: movement of a limb away from the midline

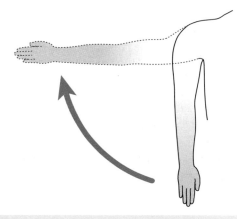

Figure 3.22   Abduction

Rotation: movement of a bone around an axis (180 degrees)

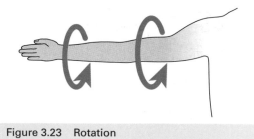

Figure 3.23   Rotation

Circumduction: a circular movement of a joint (360 degrees)

Figure 3.24   Circumduction

Supination: turning the hand so that the palm is facing upwards

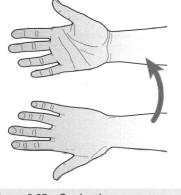

Figure 3.25   Supination

Pronation: turning the hand so that the palm is facing downwards

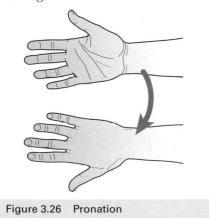

Figure 3.26   Pronation

Eversion: soles of the feet face outwards

Figure 3.27   Eversion

Inversion: soles of the feet face inwards

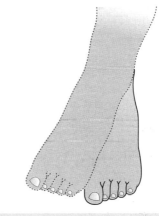

Figure 3.28   Inversion

Posture

Posture is a measure of balance and body alignment and is the maintenance of strength and tone of the body's muscles against gravity. Good posture is said to be when the maximum efficiency of the body is maintained with the minimum effort. When evaluating posture, an imaginary line is drawn vertically through the body. This is called the centre of gravity line. From the front or back this line should divide the body into two symmetrical halves. Good posture is as follows:

- with feet together the ankles and knees should touch
- hips should be the same height
- shoulders should be level
- sternum and vertebral column should run down the centre of the body in line with the centre of gravity line
- head should be erect and not tilted to one side.

Posture varies considerably in individuals and is influenced by factors such as body frame size, heredity, occupation, habits and personality. Additional factors which may also affect posture include clothing, shoes and furniture.

Good posture is important as it:

- allows a full range of movement
- improves physical appearance
- keeps muscle action to a minimum, thereby conserving energy and reducing fatigue
- reduces susceptibility to injuries
- aids the body's systems to function efficiently.

Poor posture may have the following effects on the body:

- produce alterations in body function and movement
- waste energy
- increase fatigue
- increase the risk of backache and headaches
- impair breathing
- increase the risk of muscular, ligament or joint injury
- affect circulation
- affect digestion
- give a poor physical appearance.

Postural defects

Kyphosis

This is an abnormally increased outward curvature of the thoracic spine. In this condition the back appears round as the shoulders point forwards and the head moves forwards. A tightening of the pectoral muscles is common in this condition.

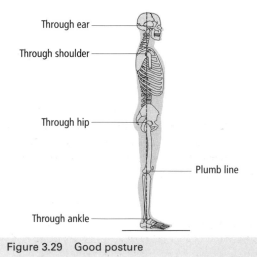

Through ear

Through shoulder

Through hip

Plumb line

Through ankle

Figure 3.29   Good posture

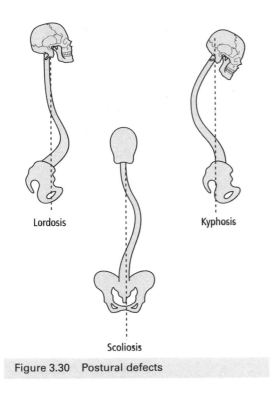

Lordosis

Kyphosis

Scoliosis

Figure 3.30 Postural defects

protrude and the knees may be hyperextended. Typical problems associated with this condition are tightening of the back muscles followed by a weakening of the abdominal muscles. Hamstring problems are common because of the anterior tilt of the pelvis. Increased weight gain or pregnancy may cause or exacerbate this condition.

Scoliosis

This is a lateral curvature of the vertebral column, either to the left or right side. Evident signs of this condition include unequal leg length, distortion of the ribcage, unequal position of the hips or shoulders and curvature of the spine (usually in the thoracic region).

Key note

Poor posture or misalignment of the body is frequently found to be the cause of continued or chronic pain as the body makes compensatory changes which are habit forming.

Lordosis

This is an abnormally increased inward curvature of the lumbar spine. In this condition the pelvis tilts forwards and as the back is hollow, the abdomen and buttocks

Disorders of the skeletal system

Ankylosing spondylitis

This is a systemic joint disease characterised by inflammation of the intervertebral disc spaces, costo-vertebral and sacroiliac joints. Fibrosis, calcification, ossification and stiffening of joints are common and the spine becomes rigid. Typically, a person will complain of persistent or intermittent lower back pain. Kyphosis is present when the thoracic or cervical regions of the spine are affected and

the weight of the head compresses the vertebrae and bends the spine forwards. This condition can cause muscular atrophy, loss of balance and falls. Typically this disease affects young male adults.

Arthritis – gout

This is a joint disorder due to deposition of excessive uric acid crystals accumulating in the joint cavity. It commonly affects the peripheral joints, often the metatarsophalangeal joint of the big toe. Kidneys can be affected. Other cartilage may be involved, including the ear pinna.

Arthritis – osteoarthritis

This is a joint disease characterised by the breakdown of articular cartilage, growth of bony spikes, swelling of the surrounding synovial membrane and stiffness and tenderness of the joint. Is also known as degenerative arthritis. It is common in the elderly and takes a progressive course. This condition involves varying degrees of joint pain, stiffness, limitation of movement, joint instability and deformity. It commonly affects the weight-bearing joints – the hips, knees, lumbar and cervical vertebrae.

Arthritis – rheumatoid

This is a chronic inflammation of peripheral joints resulting in pain, stiffness and potential damage to joints. It can cause severe disability. Joint swellings and rheumatoid nodules are tender.

Bunion

This is a swelling of the joint between the big toe and the first metatarsal. Bunions are usually caused by ill-fitting shoes and are made worse by excessive pressure.

Bursitis

This condition is the inflammation of a bursa (small sac of fibrous tissue that is lined with synovial membrane and filled with synovial fluid). It usually results from injury or infection and produces pain, stiffness and tenderness of joint adjacent to the bursa.

Dupuytren's contracture

This is the forward curvature of the fingers (usually the ring and little fingers) caused by contracture of the fibrous tissue in the palm and fingers.

Fracture

A fracture is a breakage of a bone, either complete or incomplete. There are six different types:

- A **simple fracture** involves a clean break with little damage to surrounding tissues and no break in the overlying skin (also known as a closed fracture).

- A **compound fracture** is an open fracture where the broken ends of the bone protrude through the skin.

- A **comminuted fracture** is where the bone has splintered at the site of impact and smaller fragments of bone lie between the two main fragments.

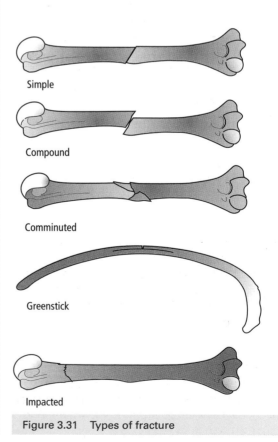

Simple

Compound

Comminuted

Greenstick

Impacted

Figure 3.31   Types of fracture

- A **greenstick fracture** only occurs in children and is a partial fracture in which one side of the bone is broken and the other side bends.

- An **impacted fracture** is where one fragment of bone is driven into the other.

- A **complicated fracture** occurs when a broken bone damages tissues and/or organs around it.

Frozen shoulder (adhesive capsulitis)

This chronic condition causes pain and stiffness and reduced mobility (or locking) of the shoulder joint. This may follow an injury, stroke or myocardial infarction or may develop due to incorrect lifting or a sudden movement.

Osteoporosis

This condition is caused by brittle bones due to ageing and the lack of the hormone oestrogen which affects the ability to deposit calcium in the matrix of bone. This can also result from prolonged use of steroids. Vulnerability to osteoporosis can be inherited. Bones can break easily and vertebrae can collapse.

Spina bifida

This is a congenital defect of the vertebral column in which the halves of the neural arch of a vertebra fail to fuse in the midline.

Sprain

A sprain is the injury to a ligament caused by overstretching or tearing. A sprain occurs when the attachments to a joint are stressed beyond their normal capacity, resulting in pain and swelling. The ankle joint and lower back are often sprained.

Stress

Stress can be defined as any factor which affects physical or emotional health. Examples of excessive stress on the skeletal system include poor posture, stiff joints and repetitive strain injuries.

Synovitis

This is the inflammation of a synovial membrane in a joint.

Temporomandibular joint tension (TMJ syndrome)

This is a collection of symptoms and signs produced by disorders of the temporo-mandibular joint. It is characterised by bilateral or unilateral muscle tenderness and reduced motion. It presents with a dull, aching pain around the joint, often radiating to the ear, face, neck or shoulder. The condition may start off as clicking sounds in the joint. There may be protrusion of the jaw or hypermobility and pain on opening the jaw. It slowly progresses to decreased mobility of the jaw and locking of the jaw may occur. Causes include chewing gum, biting nails, biting off large chunks of food, habitual protrusion of the jaw, tension in the muscles of the neck and back and clenching of the jaw. It may also be caused by injury and trauma to the joint or through a whiplash injury.

Whiplash

This condition is caused by damage to the muscles, ligaments, intervertebral discs or nerve tissues of the cervical region by sudden hyperextension and/or flexion of the neck. The most common cause is a road traffic accident when acceleration/deceleration causes sudden stretch of the tissue around the cervical spine. It may also occur as a result of hard impact sports. It can present with pain and limitation of neck movements with muscle tenderness, which can start hours to days after the accident and may take months to recover from.

- Joints hold bones together via **ligaments** and provide flexibility by facilitating movement.

- Structurally, joints are classified as **fibrous**, **cartilaginous** or **synovial**.

- **Fibrous** joints are immovable, such as the sutures of the skull bones.

- **Cartilaginous** joints are slightly movable, such as between the vertebrae of the spine.

- **Synovial** joints are freely movable joints and there are several different types: **ball and socket** (hip), **hinge** (knee and elbow), **condyloid** (wrist), **gliding** (between the vertebrae), **pivot** (between the first and second cervical vertebrae), **saddle** (between the trapezium and metacarpal of the thumb).

- Features of a **synovial joint** include a **joint (synovial) cavity**, a **fibrous joint capsule** and a **synovial membrane** containing **synovial fluid**.

Multiple-choice questions

1. The skeleton is made up of:

a 208 bones
b 206 bones
c 106 bones
d 80 bones

2. Which of the following statements is false?

a the skeleton provides protection of vital organs
b the skeleton produces blood cells in red bone marrow
c the skeleton stores vitamins
d the skeleton provides support for the weight of the body

3. The bone of the skull forms the upper sides and the back of the roof of the skull is the:

a sphenoid
b occipital
c parietal
d temporal

4. The largest bone in the face is the:

a zygomatic
b lacrimal
c maxilla
d turbinate

5. The vertebral column consists of how many movable bones?

a 24
b 33
c 12
d 9

6. The appendicular skeleton consists of:

a the sternum, shoulder girdle and ribs
b the shoulder girdle, bones of the lower and upper limb and the pelvic girdle
c the skull, vertebral column, sternum and ribs
d the sternum, bones of the upper and lower limb, and the shoulder girdle

7. Which of the following statements is true in relation to cartilage?

a cartilage receives a generous blood supply
b cartilage strengthens body structures
c cartilage is completely flexible
d cartilage cushions and absorbs shock

8. The type of joint that permits free movement is:

a synovial
b fibrous
c cartilaginous
d none of the above

9. The bone forming the posterior part of the shoulder girdle is the:

a clavicle
b scapula
c sternum
d manubrium

10. The long bone of the upper arm is the:

a radius
b ulna
c humerus
d occipital

The functions of the muscular system

The muscular system consists largely of skeletal muscle tissue which covers the bones on the outside and connective tissue which attaches muscles to the bones of the skeleton. Muscles, along with connective tissue, help to give the body its contoured shape.

The muscular system has three main functions:

Movement

Consider the action of picking up a pen that has dropped on to the floor. This seemingly simple action of retrieving the pen involves the coordinated action of several muscles pulling on bones at joints to create movement. Muscles are also involved in the movement of body fluids such as blood, lymph and urine. Consider also the beating of the heart which is continuous throughout life.

Maintaining posture

Some fibres in a muscle resist movement and create slight tension in order to enable us to stand upright. This is essential since without body posture we would be unable to maintain normal body positions such as sitting down or standing up.

The production of heat

As muscles create movement in the body they generate heat as a by-product which helps to maintain our normal body temperature.

Muscle tissue

Muscle tissue makes up about 50 per cent of your total body weight and is composed of:

- 20 per cent protein
- 75 per cent water
- 5 per cent mineral salts, glycogen and fat.

There are three types of muscle tissue in the body:

- **skeletal** or **voluntary** muscle tissue which is primarily attached to bone
- **cardiac** muscle tissue which is found in the walls of the heart
- **smooth** or **involuntary** muscle tissue which is found inside the digestive and urinary tracts, as well as in the walls of blood vessels.

All three types of muscle tissue differ in their structure and functions and the degree of control the nervous system has on them.

Voluntary muscle tissue

Voluntary muscle tissue is made up of bands of elastic or contractile tissue bound together in bundles and enclosed by a connective tissue sheath which protects the muscle and helps to give it a contoured shape.

Voluntary or skeletal muscle tissue has very little intercellular tissue. It consists almost entirely of muscle fibres held together by fibrous connective tissue and penetrated by numerous tiny blood vessels and nerves. The long slender fibres that make up muscle cells vary in size. Some are around 30 cm in length, whereas others are microscopic.

Each muscle fibre is enclosed in an individual wrapping of fine connective tissue called the endomysium. These are further wrapped together in bundles, known as fasciculi, and are covered by the perimysium (fibrous sheath), and are then gathered to from the muscle belly (main part of the muscle) with its own sheath – the fascia epimysium.

The relatively inelastic parts of each of the muscles are tendons and these are usually made up of the continuation of the endomysium and perimysium.

Each muscle fibre is made up of even thinner fibres called myofibrils. These consist of long strands of microfilaments, made up of two different types of protein strands called actin and myosin. It is the arrangement of actin and myosin filaments which gives the skeletal muscle its striated or striped appearance when viewed under a microscope. Muscle fibre contraction results from a sliding movement within the myofibrils in which actin and myosin filaments merge.

Most skeletal muscles are made up of a combination of the following types of fibres:

Fast-twitch fibres (white) – these have fast, strong reactions but tire quickly. They are well adapted for rapid movements and short bursts of activity. They have a rich blood supply and mainly use the energy stores of glucose in the muscles which can be transferred into mechanical energy without oxygen.

Slow-twitch fibres (red) – these fibres have greater endurance but do not produce as much strength as fast-twitch fibres. They are therefore suited to slower and more sustained movements and are relatively resistant to fatigue. Their energy comes from the breaking down of glucose by oxygen and they depend on a continuous supply of glucose for endurance. Slow-twitch fibres have good circulation (the red colour comes from both the circulation and the presence of a red pigmented protein that stores the oxygen).

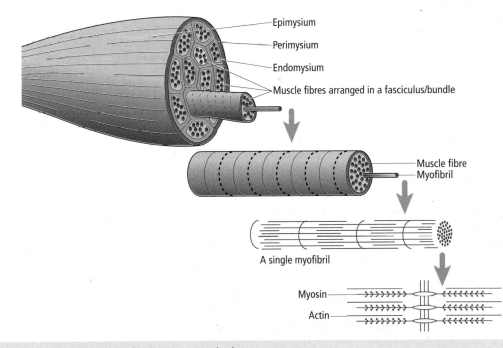

Figure 4.1   The structure of voluntary muscle tissue

During low-intensity work, such as walking, the body is working well below its maximal capacity and only slow-twitch fibres are working. As muscle intensity increases and the exercise becomes more anaerobic, fast-twitch fibres are activated. Whatever the intensity of movement only a small number of fibres are used at any one time to prevent damage and injury to the tissues.

Key note

Each person is born with a set number of muscle fibres which cannot be increased. An increase in the size of a muscle is due to exercise which will cause an increase in the individual fibres. However, with disuse these will shrink again as the muscle atrophies. It is interesting to note that men are more able to enlarge their muscles through exercise than women, due to the effects of male hormones.

The way in which the bundles of fibres lie next to one another in a muscle will determine its shape. The contractile force of a muscle is partly attributable to the architecture of its fibres. Common muscle fibre arrangements include:

Parallel – muscles with parallel fibres can vary from short, flat muscles to spindle-shaped (fusiform) muscles to long straps.

Convergent – this is where the muscle fibres converge towards a single point for maximum concentration of the contraction. The direction of movement is determined by which sections of the muscle are activated. The muscle may be a triangular sheet (the pectoralis major muscle or the latissimus dorsi). These muscles often cross joints that have a large range of possible movements.

They provide a strong but steady pull, fine-tuning the angle of movement and thus balancing movement with continuing stability in the joint.

Pennate – this is where the fibres lie at an angle to the tendon and, therefore, also to the direction of pull. They have lots of short fibres so the muscle pull is short but strong. They may be further classified as follows:

- **Unipennate** – diagonal fibres attach to one side of the tendon only, such as the soleus.

- **Bipennate** – the fibres converge on to a central tendon from both sides, such as the rectus femoris.

- **Multipennate** – the muscle has several tendons of origin, such as the deltoid.

Key note

Every time a muscle is used, the muscle fibres shorten along their length, therefore tension often accumulates in lines in the longer muscles (particularly those with parallel fibres such as the paravertebral muscles). In muscles with shorter fibre pennates and convergent fibres tension is often in knots rather than in lines.

Voluntary muscle works intimately with the nervous system and, therefore, will only contract if a stimulus is applied to it via a motor nerve. Each muscle fibre receives its own nerve impulse so that fine and varied motions are possible. Voluntary muscles also have their own small, stored supply of glycogen which is used as fuel for energy. Voluntary muscle tissue differs from other types of muscle tissue in that the muscles tire easily and need regular exercise.

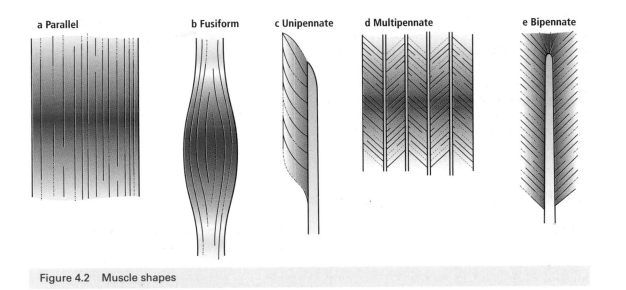

Figure 4.2   Muscle shapes

Cardiac muscle

Cardiac muscle is a specialised type of involuntary muscle tissue found only in the walls of the heart. Forming the bulk of the wall of each heart chamber, cardiac muscle contracts rhythmically and continuously to provide the pumping action necessary to maintain a relatively consistent flow of blood throughout the body. Cardiac muscle resembles voluntary muscle tissue in that it is striated due to the actin and myosin filaments. However, it differs in two ways:

■ It is branched in structure.

■ It has intercalated discs in between each cardiac muscle cell which form strong junctions to assist in the rapid transmission of impulses throughout an entire section of the heart, rather than in bundles.

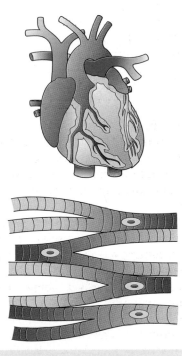

Figure 4.3   Cardiac muscle tissue

The contraction of the heart is automatic. The stimulus to contract comes from a specialised area of muscle in the heart called the sinoatrial (SA) node which controls the heart rate.

As the heart has to alter its force of contraction due to regional requirements, its contraction is regulated not only by nerves but also by hormones, such as adrenaline in the blood, which can speed up contractions.

Smooth muscle

Smooth muscle is also known as involuntary muscle, as it is not under the control of the conscious part of the brain. It is found in the walls of hollow organs such as the stomach, intestines, bladder, uterus and in blood vessels.

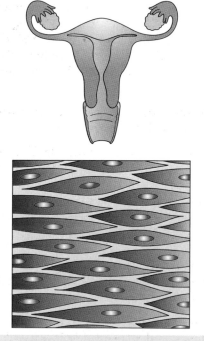

Figure 4.4   Smooth/involuntary muscle tissue

The main characteristics of smooth muscle are:

■ The muscle cells are spindle-shaped and tapered at both ends.

■ Each muscle cell contains one centrally located, oval-shaped nucleus.

Smooth muscle has no striations due to the different arrangement of the protein filaments actin and myosin, which are attached at their ends to the cell's plasma membrane.

The muscle fibres of smooth muscle are adapted for long, sustained contraction and, therefore, consume very little energy. One of the special features of smooth muscle is that it can stretch and shorten to a greater extent and still maintain its contractile function. Smooth muscle will contract or relax in response to nerve impulses, stretching or hormones, but it is not under voluntary control.

Smooth muscle, like voluntary muscle, has muscle tone and this is important in areas such as the intestines where the walls have to maintain a steady pressure on the contents.

Muscle contraction

Muscle tissue has several characteristics which help contribute to the functioning of a muscle:

■ **Contractibility** – the capacity of the muscle to shorten and thicken

■ **Extensibility** – the ability to stretch when the muscle fibres relax

■ **Elasticity** – the ability to return to its original shape after contraction

■ **Irritability** – the response to stimuli provided by nerve impulses.

Muscles vary in the speed at which they contract. The muscle in your eyes will be moving very fast as you are reading this page, while the muscles in your limbs assisting you in turning the pages will be contracting at a moderate speed. The speed of a muscle contraction, therefore, is modified to meet the demands of the action concerned and the degree of nervous stimulus it has received.

Stimulus to contract

Skeletal or voluntary muscles are moved as a result of nervous stimulus which they receive from the brain via a motor nerve. Each skeletal fibre is connected to the fibre of a nerve cell. Each nerve fibre ends in a motor point which is the end portion of the nerve and is the part through which the stimulus to contract is given to the muscle fibre. A single motor nerve may transmit stimuli to one muscle fibre or as many as 150, depending on the effect of the action required.

The site where the nerve fibre and muscle fibre meet is called a neuromuscular junction. In response to a nerve impulse, the end of the motor nerve fibre secretes a neurotransmitter substance called acetylcholine, which diffuses across the junction and stimulates the muscle fibre to contract.

Cardiac and smooth muscle are innervated by the autonomic nervous system.

The contraction of voluntary muscle tissue

The functional characteristic of muscle is its ability to transform chemical energy into mechanical energy in order to exert force. Muscles exert force by contracting or making themselves shorter.

The role of actin and myosin

A voluntary or skeletal muscle consists of many long, cylindrical fibres. Each of these fibres is in turn filled with long bundles of even smaller fibres called myofibrils. A myofibril resembles stacked blocks. In each block (or sarcomere) thick filaments containing the protein myosin overlap thin filaments containing the protein actin. Sarcomeres are divided into dark Z lines, with their centres known as H zones. As the muscle contracts, its sarcomeres shorten, reducing the distance between the Z lines and the width of the H zone. Muscle fibre contraction results from a sliding movement within the myofibrils in which the actin and myosin filaments merge. Actin and myosin affect contraction in the following ways:

■ During contraction a sliding movement occurs within the contractile fibres (myofibrils) of the muscle in which the actin protein filaments move inwards towards the myosin and the two filaments merge.

■ Cross-bridges of myosin filaments form linkages with actin filaments.

■ This action causes the muscle fibres to shorten and thicken and then pull on their attachments (bones and joints) to effect the movement required.

■ The attachment of myosin cross-bridges to actin requires the mineral calcium.

■ The nerve impulses leading to contraction cause an increase in calcium ions within the muscle cell.

■ During relaxation the muscle fibres elongate and return to their original shape.

The force of muscle contraction depends on the number of fibres in a muscle which contract simultaneously. The more fibres involved, the stronger and more powerful the contraction will be.

Key note

The basic contractile process is the same in cardiac, smooth and voluntary muscles, with movement being achieved through the action of the protein filaments actin and myosin. However, since the requirements are different in terms of speed and force of contraction, the structure of cardiac and smooth muscles are slightly different to voluntary muscle tissue.

The energy needed for muscle contraction

A certain amount of energy is needed to effect the mechanical action of the muscle fibres. This is obtained principally from carbohydrate foods such as glucose in the arterial blood supply. Glucose, which is not required immediately by the body, is converted into glycogen and stored in the liver and the muscles. Muscle glycogen, therefore, provides the fuel for muscle contraction. The process is as follows:

■ During muscle contraction glycogen is broken down by a process called oxidation, where glucose combines with oxygen and releases energy. Oxygen is stored in the form of haemoglobin in the red blood cells and as myoglobin in the muscle cells.

■ During oxidation, a chemical compound called ATP (adenosine triphosphate) is formed. Molecules of ATP are contained within voluntary muscle tissue and their function is to temporarily store energy produced from food.

■ When the muscle is stimulated to contract, ATP is converted to another chemical compound, ADP (adenosine diphosphate), which releases the energy needed to be used during the phase of muscle contraction.

■ During the oxidation of glycogen, a substance called pyruvic acid is formed.

■ If plenty of oxygen is available to the body, as in rest or undertaking moderate exercise, then the pyruvic acid is broken down into waste products, carbon dioxide and water, which are excreted into the venous system. This is known as aerobic respiration.

■ If insufficient oxygen is available to the body, as may be the case with vigorous exercise, then the pyruvic acid is converted into lactic acid. This is known as anaerobic respiration.

Key note

The waste product lactic acid, which diffuses into the bloodstream after vigorous exercise, causes the muscles to ache. This condition is known as muscle fatigue, which is defined as the loss of the ability of a muscle to contract efficiently due to insufficient oxygen, exhaustion of energy supply and the accumulation of lactic acid.

The effects of increased circulation on muscle contraction

During exercise muscles require more oxygen to cope with the increased demands made on the body. The body is then active in initiating certain circulatory and respiratory changes to meet the increased oxygen requirements of the muscles.

Circulatory changes that occur in the body during muscle contraction

During exercise there is an increased return of venous blood to the heart, owing to the more extensive movements of the diaphragm and the general contractions of the muscles compressing the veins. With the rate and output of each heartbeat being increased, a greater volume of blood is circulated around the body, which leads to an increase in the amount of oxygen in the blood.

More blood is distributed to the muscle and less to the intestine and skin to meet the needs of the exercising muscles. During exercise a muscle may receive as much as 15 times its normal flow of blood.

Respiratory changes

The presence of lactic acid in the blood stimulates the respiratory centre in the brain, increasing the rate and depth of breath, producing panting. The rate and depth of breath remains above normal for a while after strenuous exercise has ceased. Large amounts of oxygen are taken in to allow the cells of the muscles and the liver to dispose of the accumulated lactic acid by oxidising it and converting it to glucose or glycogen. Lactic acid is formed in the tissues in amounts far greater than can be disposed of immediately by available oxygen. The extra oxygen needed to remove the accumulated lactic acid is what is called the oxygen debt, which must be repaid after the exercise is over.

Key note

The conversion of lactic acid back into glucose is a relatively slow process and it may take several hours to repay the oxygen debt, depending on the extent of the exercise undertaken. This situation can be minimised by massaging muscles before and after an exercise schedule which will increase the blood supply to the muscles and prevent an excess of lactic acid forming in the muscles.

The effects of temperature on muscle contraction

Exercising muscles produces heat, which is carried away from the muscle by the bloodstream and is distributed to the rest of the body. Exercise is, therefore, an effective way to increase body temperature. When muscle tissue is warm, the process of contraction will occur faster due to the acceleration of the chemical reactions and the increase in circulation. However, it is possible for heat cramps to occur in muscles which are exercised at high temperatures, as increased sweating causes loss of sodium in the body, leading to a reduction in the concentration of sodium ions in the blood supplying the muscle.

Cramp occurs when muscles become over-contracted and hence go into spasm. This is usually caused by an irritated nerve or an imbalance of mineral salts such as sodium in the body. Cramp most commonly affects the calf muscles or the soles of the feet. Cramp can be very painful as it is a sudden involuntary contraction of the muscle.

Treatment to relieve the pain of cramp includes stretching the affected muscle group and using soothing effleurage movements to help to relax the muscles. Conversely, as muscle tissue is cooled, the chemical reactions and circulation slow, causing the contraction to be slower. This causes an involuntary increase in muscle tone known as shivering, which increases body temperature in response to cold.

Muscle tone

Even in a relaxed muscle a few muscle fibres remain contracted to give the muscle a certain degree of firmness. At any given time a small number of motor units in a muscle are stimulated to contract and cause tension in the muscle rather than full contraction and movement, while the others remain relaxed. The group of motor units functioning in this way change periodically so that muscle tone is maintained without fatigue. This state of partial contraction of a muscle is known as muscle tone and is important for maintaining body posture.

Good muscle tone may be recognised by the muscles appearing firm and rounded, whereas poor muscle tone may be recognised by the muscles appearing loose and flattened.

Key note

An increase in the size and diameter of muscle fibres, usually caused by exercise and weight lifting, leads to a condition called hypertrophy.

Muscles with less than the normal degree of tone are said to be flaccid, and when the muscle tone is greater than normal the muscles become spastic and rigid.

Key note

Muscle tone will vary from person to person and will depend largely on the amount of exercise undertaken. Muscles with good tone have a better blood supply as their blood vessels will not be inhibited by fat.

Muscle attachments

In order to understand how skeletal muscles produce movement, it is helpful first to understand how muscles are attached to the rest of the body.

Tendons

Tendons are tightly woven, white, glistening, tough fibrous bands or cords that link muscle to bone. They do not stretch or contract the way muscles do and, therefore, are not at all elastic.

A tendon's blood supply is limited so it usually does not heal quickly or easily. Tendons are mechanically strong, as their primary role is to transmit the contractile force of the muscle to the bone. For this reason, tendons are relatively inflexible structures, designed to be strongest in the direction of tensile stress.

Despite their great strength, tendons are most susceptible to excessive tensile stress injuries. Luckily, complete tendon tears or ruptures are infrequent (the most common ruptured tendon is the Achilles tendon).

Ligaments

Ligaments are strong, fibrous, elastic tissues that are usually cordlike in nature. They are placed parallel with or closely interlaced with one another, which creates a white, shining, silvery effect. A ligament is pliant so as to allow good freedom of movement, but is also strong, tough and inextensible (does not stretch).

Their attachments to various skeletal components help to maintain the bones in correct relationship to one another, stabilising the joints. When torn, ligaments heal slowly due to the fact that they have a relatively poor blood supply compared to muscles and tendons.

The orientation of a ligament's fibres (parallel arrangements complemented by transverse fibres) gives the ligament an ability to resist stress in several different planes. Ligaments also contain a greater concentration of elastin than a tendon. This will allow the ligament a small degree of 'give' before it pulls taut at a particular joint. This small amount of 'give' is important because if ligaments were as rigid and 'ungiving' to tensile stress as a tendon, the frequency of ligament injuries would be much greater. The most common injury to a ligament is a sprain. Ligamentous tearing is generally referred to as a sprain.

Fasciae

Fasciae consist of fibrous connective tissue that envelops certain muscles which then forms partitions for others. Fascia is all encompassing. It packages, supports and envelops all the body's muscles and organs. It separates different muscles yet allows them to glide smoothly beside each other. The fascial planes provide pathways for nerves, blood vessels and lymphatic vessels.

Fasciae, therefore, play a key role in maintaining the health of a muscle. When these cellophane-like sheets become adhered to neighbouring muscle tissue, efficiency and function can be significantly diminished. If the fasciae becomes torn or overstressed, their subsequent loss of elasticity will cause and maintain chronic tissue congestion.

Key note

One of the most problematic features of fascia is its response to prolonged immobilisation. If the body Is held in one position for long periods of time, the fascia has a tendency to adapt to that position. This is especially problematic when the fascia is held in a shortened position. When it is kept in this shortened position, it will structurally adapt to that position and resist an attempt to return to its normal length.

Origins and insertions

Muscle attachments are known by the terms origin and insertion. Generally, the end of the muscle closest to the centre of the body is referred to as the origin, and the insertion is the furthest attachment.

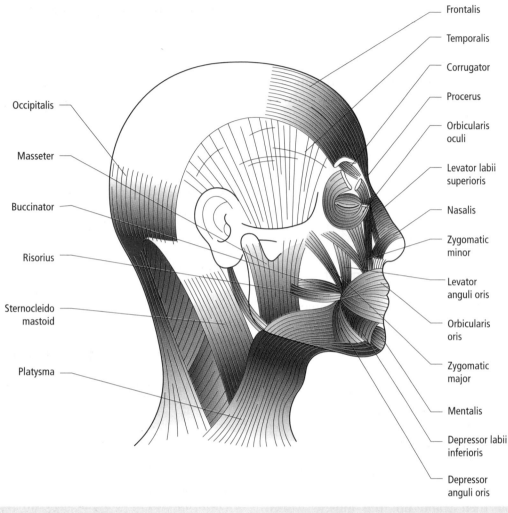

Frontalis

Temporalis

Corrugator

Procerus

Orbicularis oculi

Levator labii superioris

Nasalis

Zygomatic minor

Levator anguli oris

Orbicularis oris

Zygomatic major

Mentalis

Depressor labii inferioris

Depressor anguli oris

Occipitalis

Masseter

Buccinator

Risorius

Sternocleido mastoid

Platysma

Figure 4.5   Muscles of the head and neck

Key note

The temporalis muscle becomes over-tight and painful in the condition known as temporomandibular joint syndrome. The temporalis muscle also becomes tightened with a tension headache.

Orbicularis oculi

- **Position and attachments** – this is a circular muscle that surrounds the eye. It is attached to the bones at the outer edge and the skin of the upper and lower eyelids at the inner edge.

- **Action** – closes the eye.

Key note

The orbicularis oculi muscle is used when blinking or winking. It also compresses the lacrimal gland, aiding the flow of tears.

Orbicularis Oris

- **Position and attachments** – this is a circular muscle that surrounds the mouth. Its fibres attach to the maxillae, mandible, lips and buccinator muscle.
- **Action** – closes the mouth.

Key note

The orbicularis oris muscle is used when shaping the lips for speech and when kissing. It also contracts the lips when tense.

Corrugator

- **Position and attachments** – this muscle is located in between the eyebrows. It is attached to the frontalis muscle and the inner edge of the eyebrow.
- **Action** – brings the eyebrows together.

Key note

The corrugator muscle is used when frowning.

Procerus

- **Position and attachments** – this muscle is located in between the eyebrows. It is attached to the nasal bones and the frontalis muscle.

- **Action** – draws the eyebrows inwards.

Key note

The contraction of the procerus muscle creates a puzzled expression.

Nasalis

- **Position and attachments** – this muscle is located at the sides of the nose. It is attached to the maxillae bones and the nostrils.
- **Action** – dilates and compresses the nostrils.

Key note

The nasalis muscle is used when blowing the nose.

Zygomatic major and minor/zygomaticus

- **Position and attachments** – lies in the cheek area, extending from the zygomatic bone to the angle of the mouth.
- **Action** – draws the angle of the mouth upwards and laterally.

Key note

The zygomaticus muscle is used when laughing or smiling.

Levator labii superioris

- **Position and attachments** – this muscle is located towards the inner cheek, beside the nose, and extends from the upper jaw to the skin of the corners of the mouth and the upper lip.
- **Action** – raises the upper lip and the corners of the mouth.

Key note

The levator labii superioris muscle is used to create a snarling expression.

Levator anguli oris

- **Position and attachments** – this muscle extends from the maxilla (upper jaw) to the angle of the mouth.
- **Action** – elevation of the angle of the mouth.

Key note

The levator anguli oris is also known as the caninus (kay-**ni**-nus), as its contraction can result in the teeth, especially the canine tooth, becoming visible.

Depressor anguli oris

- **Position and attachments** – this muscle extends from the mandible (lower jaw) to the angle of the mouth.
- **Action** – depression of the angle of the mouth.

Key note

The action of the depressor anguli oris contributes to the facial expression of sadness or uncertainty.

Depressor labii inferioris

- **Position and attachments** – this muscle extends from the mandible to the midline of the lower lip.
- **Action** – depression of the lower lip.

Key note

The actions of the depressor labii inferioris can contribute to the facial expressions of sorrow, doubt or irony.

Lateral pterygoid

- **Position and attachments** – this muscle extends from the sphenoid bone to the mandible and temporomandibular joint.
- **Action** – protraction of the mandible

Key note

Tension in the lateral pterygoid may be associated with dysfunction of the temporomandibular joint (TMJ syndrome).

Medial pterygoid

- **Position and attachments** – this muscle extends from the sphenoid bone to the internal surface of the mandible.
- **Action** – elevation of the mandible.

Key note

Tension in the medial pterygoid may be associated with dysfunction of the temporomandibular joint (TMJ syndrome).

Risorius

- **Position and attachments** – this is a triangular-shaped muscle that lies horizontally on the cheek, joining at the corners of the mouth. The risorius lies above the buccinator muscles (see below) and is attached to the zygomatic bone at one end and the skin of the corner of the mouth at the other.

- **Action** – pulls the corners of the mouth sideways and upwards.

Key note

The risorius muscle creates a grinning expression.

Buccinator

- **Position and attachments** – this muscle is the main muscle of the cheek. It is attached to both the upper and lower jaws. Its fibres are directed forwards from the bones of the jaws to the angle of the mouth.

- **Action** – this muscle helps hold food in contact with the teeth when chewing and compresses the cheek.

Key note

The buccinator muscle is used when blowing up a balloon or blowing a trumpet. It is also a common site for holding tension in the face.

Mentalis

- **Position and attachments** – this muscle radiates from the lower lip over the centre of the chin. It is attached to the lower jaw and the skin of the lower lip.

- **Action** – elevates the lower lip and wrinkles the skin of the chin.

Key note

The mentalis muscle is used when expressing displeasure and when pouting.

Masseter

- **Position and attachments** – this is a thick, flattened and superficial muscle. Its fibres extend downwards from the zygomatic arch to the mandible.

- **Action** – this is the main muscle of mastication. It raises the jaw and exerts pressure on the teeth when chewing.

Key note

The masseter muscle is a facial muscle that tends to hold a lot of tension and can be felt just in front of the ear when the teeth are clenched.

Sternocleidomastoid

- **Position and attachments** – this is a long muscle that lies obliquely across each side of the neck. Its fibres extend upwards from the sternum and clavicle at one end to the mastoid process of the temporal bone (at the back of the ear).

- **Action** – when working together they flex the neck (pull the chin down towards the chest), and when working individually they rotate the head to the opposite side.

Key note

Spasm of the sternocleidomastoid muscle results in a condition known as torticollis or wryneck. Sternocleido-mastoid is the only muscle that moves the head but is not attached to any vertebrae.

Platysma

- **Position and attachments** – this is a superficial neck muscle that extends from the chest (fasciae covering the upper part of pectoralis major and deltoid), up either side of the neck to the chin.

- **Action** – depresses the lower jaw and lower lip.

Key note

The platysma muscle is used in yawning and when creating a pouting expression.

Splenius capitis

- **Position and attachments** – this is a long, posterior neck muscle that extends from the spinous processes of C7–T3 to the mastoid process of the temporal bone and the occipital bone.

- **Action** – extension of the head and neck, and lateral flexion of the head and neck.

Key note

The splenius capitis muscle is shaped like a bandage and attaches to the head. The right and left splenius capitis muscles form a 'V' shape and because of this they are sometimes referred to as the 'golf tee' muscles.

Splenius cervicis

- **Position and attachments** – this is a long, posterior neck muscle (fibres slightly thinner and longer than splenius capitis) that extends from the spinous processes of T3–T6 to the transverse processes of C1–C3.

- **Action** – extension of the neck, and lateral flexion of the neck.

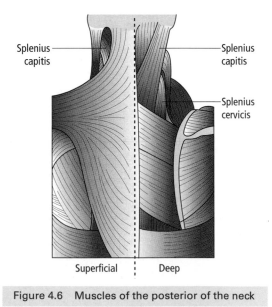

Figure 4.6 Muscles of the posterior of the neck

Key note

The splenius cervicis muscle is shaped like a bandage and attaches to the cervical spine (the neck). The right and left splenius cervicis muscles form a 'V' shape, and because of this they are sometimes referred to as the 'golf tee' muscles.

The muscles of the shoulder

Trapezius

- **Position and attachments** – this is a large, triangular-shaped muscle in the upper back that extends horizontally from the base of the skull (occipital bone) and the cervical and thoracic vertebrae to the scapula. Its fibres are arranged in three groups – upper, middle and lower.

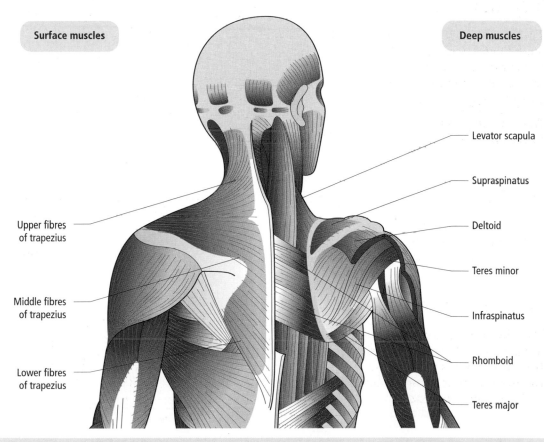

Surface muscles

Deep muscles

Levator scapula

Supraspinatus

Deltoid

Teres minor

Infraspinatus

Rhomboid

Teres major

Upper fibres of trapezius

Middle fibres of trapezius

Lower fibres of trapezius

Figure 4.7   Muscles of the shoulder

- **Action** – The upper fibres raise the shoulder girdle, the middle fibres pull the scapula towards the vertebral column and the lower fibres draw the scapula and shoulder downward. When the trapezius is fixed in position by other muscles, it can pull the head backwards or to one side.

Key note

The trapezius muscle is one of the most commonly found muscles to hold upper body tension, causing discomfort and restrictions in the neck and shoulder. Holding a bag on one shoulder can cause tightness in the upper fibres of the trapezius.

Levator scapula

- **Position and attachments** – this is a straplike muscle that runs almost vertically through the neck, connecting the cervical vertebrae to the scapula.
- **Action** – elevates and adducts the scapula.

Key note

Due to its attachments, tension in the levator scapula can affect mobility of both the neck and the shoulder.

Rhomboids

- **Position and attachments** – the fibres of these muscles lie between the scapulae. They attach to the upper thoracic vertebrae at one end and the medial border of the scapula at the other end.
- **Action** – adduct the scapula.

Key note

The rhomboid muscles are also known as the 'Christmas tree' muscles because their fibres are arranged obliquely like Christmas tree branches. Tension in the rhomboid muscles often results in aching and soreness in between the scapulae.

Supraspinatus

- **Position and attachments** – this muscle is located in the depression above the spine of the scapula. It is attached to the spine of the scapula at one end and the humerus at the other.
- **Action** – abducts the humerus, assisting the deltoid.

Key note

The supraspinatus is one of the four rotator cuff muscles (along with infraspinatus, teres major and subscapularis). The supraspinatus is the only muscle of the rotator cuff that does not rotate the humerus.

Infraspinatus

- **Position and attachments** – this muscle attaches to the middle two-thirds of the scapula below the spine of the scapula at one end and the top of the humerus at the other.
- **Action** – rotates the humerus laterally (outwards).

Key note

The infraspinatus is one of the four rotator cuff muscles (along with supraspinatus, teres minor and subscapularis). Tension in the infraspinatus muscle can affect the range of mobility in the arm and the shoulder.

Teres major

- **Position and attachments** – this muscle attaches to the bottom lateral edge of the scapula at one end and the back of the humerus (just below the shoulder joint) at the other.
- **Action** – adducts and medially (inwardly) rotates the humerus.

Key note

The teres major is sometimes referred to as the 'little helper' of the latissimus dorsi muscle because they run together between the scapula and the humerus.

Teres minor

- **Position and attachments** – this muscle attaches to the lateral edge of the scapula, above teres major at one end, and into the top of the posterior of the humerus at the other.
- **Action** – rotates the humerus laterally (outwards).

Key note

The teres minor is one of the four rotator cuff muscles (along with supraspinatus, infraspinatus and subscapularis).

Key note

Although the teres major and minor may appear similar by name, they wrap around the humerus in opposite directions and, therefore, have opposite rotary actions.

Subscapularis

- **Position and attachments** – this muscle attaches the inside surface of the scapula to the anterior of the top of the humerus.
- **Action** – rotates the humerus medially, draws the humerus forwards and down when the arm is raised.

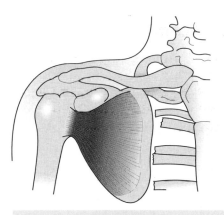

Figure 4.8   Subscapularis

Key note

The subscapularis is one of the four rotator cuff muscles (along with supraspinatus, infraspinatus and teres minor). The subscapularis muscle is often implicated in the case of a frozen shoulder.

Deltoid

- **Position and attachments** – this muscle is a thick, triangular muscle that caps the top of the humerus and the shoulder. It attaches to the clavicle and the spine of the scapula at one end, and to the side of the humerus at the other.

- **Action** – abducts the arm and draws the arm backwards and forwards.

Key note

The deltoid has anterior, lateral and posterior fibres and these give the shoulder its characteristic shape.

The muscles of the upper limbs

Coracobrachialis

- **Position and attachments** – extends from the scapula to the middle of the humerus along its medial surface.

- **Action** – flexes and adducts the humerus.

Key note

The name coracobrachialis indicates that this muscle is related to the coracoid process (in the scapula) and the brachium (the arm).

Biceps

- **Position and attachments** – anterior of the upper arm (humerus). It attaches to the scapula at one end and the radius and flexor muscles of the forearm at the other.

- **Action** – flexes the forearm at the elbow and supinates the forearm.

Key note

The actions of the biceps muscle are likened to the action of removing a corkscrew from a wine bottle.

Triceps

- **Position and attachments** – this is a posterior muscle of the upper arm (humerus). It attaches to the posterior of the humerus and outer edge of the scapula at one end, and to the ulna below the elbow at the other.

- **Action** – extension (straightening) of the forearm.

Key note

The triceps muscle is also referred to as the 'boxer's muscle' as it is used when delivering a knockout punch.

Brachialis

- **Position and attachments** – this muscle lies beneath biceps. It attaches to the distal half of the anterior surface of the humerus at one end and the ulna at the other.

- **Action** – flexes the forearm at the elbow.

Key note

The brachialis is a strong and fairly large muscle, which accounts for much of the contour of the biceps muscle.

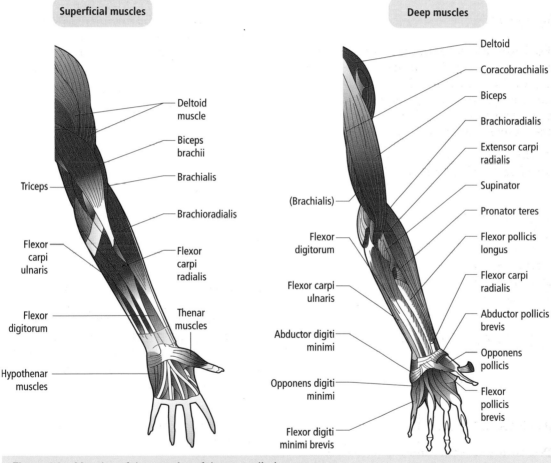

Superficial muscles

- Deltoid muscle
- Biceps brachii
- Brachialis
- Triceps
- Brachioradialis
- Flexor carpi ulnaris
- Flexor carpi radialis
- Flexor digitorum
- Thenar muscles
- Hypothenar muscles

Deep muscles

- Deltoid
- Coracobrachialis
- Biceps
- Brachioradialis
- Extensor carpi radialis
- Supinator
- (Brachialis)
- Pronator teres
- Flexor digitorum
- Flexor pollicis longus
- Flexor carpi ulnaris
- Flexor carpi radialis
- Abductor digiti minimi
- Abductor pollicis brevis
- Opponens digiti minimi
- Opponens pollicis
- Flexor pollicis brevis
- Flexor digiti minimi brevis

Figure 4.9   Muscles of the anterior of the upper limb

Pronator teres

- **Position and attachments** – this muscle crosses the anterior aspect of the elbow. It attaches to the distal end of the humerus and the upper aspect of the ulna at one end, and the lateral surface of the radius at the other.
- **Action** – pronates and flexes the forearm.

Key note

Due to the fact that the fibres of the pronator teres cross the elbow joint, irritation and inflammation of this muscle may contribute to the condition known as tennis elbow.

Supinator

- **Position and attachments** – attaches to the lateral aspect of the lower humerus and the radius.
- **Action** – supinates the forearm.

Key note

Due to the fact that the fibres of the supinator cross the elbow joint, irritation and inflammation of this muscle may contribute to the condition known as tennis elbow.

Flexors of the forearm

Brachioradialis

- **Position and attachments** – this muscle connects the humerus to the radius. It attaches to the distal end of the humerus at one end and the radius at the other end.
- **Action** – flexes the forearm at the elbow.

Key note

The brachioradialis can be felt as the bulge on the radial side of the forearm. The brachioradialis is sometimes nicknamed the 'hitchhiker muscle' for its characteristic action of flexing the forearm in a position halfway between full pronation and full supination.

Flexor carpi radialis

- **Position and attachments** – this muscle extends along the radial side of the anterior of the forearm, from the medial end of the humerus to the radial side of forearm and the base of the second and third metacarpals.
- **Action** – flexion of the wrist.

Flexor carpi ulnaris

- **Position and attachments** – this muscle extends along the ulnar side of the anterior of the forearm, from the medial end of the humerus to the pisiform and hamate carpal bones and the base of the fifth metacarpal.
- **Action** – flexion of the wrist.

Flexor carpi digitorum

- **Position and attachments** – this muscle extends from the medial end of the humerus, the anterior of the ulna and radius, to the anterior surfaces of the second to fifth fingers.
- **Action** – flexion of the fingers.

Key note

Any of the flexor muscles in the forearm can become easily inflamed due to excess pressure and overwork, a common example being working on a keyboard for extended periods of time.

Extensors of the forearm

Extensor carpi radialis

- **Position and attachments** – this muscle extends along the radial side of the posterior of the forearm, from above the lateral end of the humerus to the posterior of the base of the second metacarpal.
- **Action** – extension of the wrist.

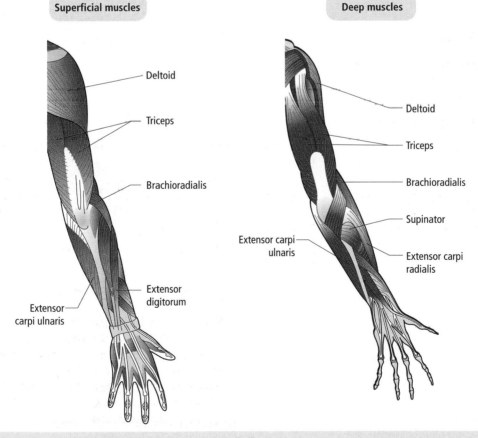

Superficial muscles

- Deltoid
- Triceps
- Brachioradialis
- Extensor digitorum
- Extensor carpi ulnaris

Deep muscles

- Deltoid
- Triceps
- Brachioradialis
- Supinator
- Extensor carpi radialis
- Extensor carpi ulnaris

Figure 4.10   Muscles of the posterior of the upper limb

Extensor carpi ulnaris

- **Position and attachments** – this muscle extends along the ulnar side of the posterior of the forearm, from above the lateral end of the humerus to the ulna and the posterior side of the base of the fifth metacarpal.
- **Action** – extension of the wrist.

Extensor digitorum

- **Position and attachments** – this muscle extends along the lateral side of the posterior of the forearm from the lateral end of the humerus to the second and fifth phalanges.
- **Action** – extension of the fingers.

Key note

Any of the extensor muscles in the forearm can become easily inflamed due to excess pressure and overwork, a common example being working on a keyboard for extended periods of time.

The muscles of the hand

Thenar eminence

This is an eminence of soft tissue located on the radial side of the palm of the hand. There are three muscles of the thenar eminence:

- abductor pollicis brevis
- flexor pollicis brevis

- opponens pollicis.

Action – all three muscles move the thumb.

Hypothenar eminence

This is an eminence of soft tissue located on the ulnar side of the palm of the hand. There are three muscles of the hypothenar eminence:

- abductor digiti minimi manus
- flexor digiti minimi manus
- opponens digiti minimi.

Action – all three muscles move the little finger.

The muscles of the lower limb

Quadriceps extensor

The quadriceps is made up of four muscles: rectus femoris, vastus lateralis, vastus intermedius and vastus medialis.

- **Position and attachments** – anterior aspect of the thigh attached to the pelvic girdle (rectus femoris) and femur (vastus group) at one end of the patella, and the tibia at the other end.
- **Action** – as a group they extend the knee and flex the hip.

Key note

The quadriceps are a group of strong muscles that are used for walking, kicking and raising the body from a sitting or squatting position.

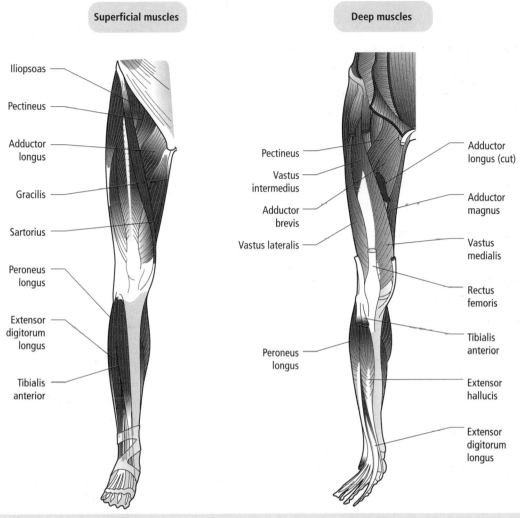

Superficial muscles

Iliopsoas
Pectineus
Adductor longus
Gracilis
Sartorius
Peroneus longus
Extensor digitorum longus
Tibialis anterior

Deep muscles

Pectineus
Vastus intermedius
Adductor brevis
Vastus lateralis
Peroneus longus

Adductor longus (cut)
Adductor magnus
Vastus medialis
Rectus femoris
Tibialis anterior
Extensor hallucis
Extensor digitorum longus

Figure 4.11   Muscles of the anterior of the lower limb

Sartorius

- **Position and attachments** – this muscle is attached to the ilium of the pelvis. It crosses the anterior of the thigh to the medial aspect of the tibia.

- **Action** – flexes the hip and knee and rotates the thigh laterally (turns it outwards).

Key note

Due to its unusual position, the sartorius can flex both the hip and the knee. Overcontraction of the sartorius can lead to knee problems because turning the leg outwards puts pressure on the knee. The sartorius is also the longest muscle in the human body.

Adductors

This is a group of four muscles: adductor brevis, adductor longus, adductor magnus and pectineus.

- **Position and attachments** – these muscles are situated on the medial aspect of the thigh. They are attached to the lower part of the pelvic girdle at one end (pubic bones and the ischium) and the inside of the femur at the other end.

- **Action** – as a group they adduct and laterally rotate the thigh. They also flex the hip.

Key note

The adductors are important muscles in the maintenance of posture. Groin strains are common problems associated with these muscles.

Gracilis

- **Position and attachments** – this is a long, straplike muscle attached to the lower edge of the pubic bone at one end and the upper part of the medial aspect of the tibia at the other end.

- **Action** – adducts the thigh, flexes the knee and hip, medially (inwardly) rotates the thigh and tibia.

Key note

The gracilis muscle is the second longest muscle in the human body.

Hamstrings

The hamstrings consist of three muscles – two situated on the inside of the thigh (semitendinosus and semimembranous) and one on the outside of the thigh (biceps femoris).

- **Position and attachments** – the posterior aspect of the thigh attaches to the lower part of the pelvis (ischium), and the lower part of the posterior of the femur to either side of the posterior of the tibia.

- **Action** – flex the knee and extend the hip.

Key note

The hamstring muscles contract powerfully when raising the body from a stooped position and when climbing stairs.

Tensor fascia lata

- **Position and attachments** – this muscle runs laterally down the side of the thigh. It is attached to the outer edge of the ilium of the pelvis and runs via the long fascia lata tendon to the lateral aspect of the top of the tibia.

- **Action** – flexes, abducts and medially rotates the thigh.

Key note

The tensor fascia lata is attached to a broad sheet of connective tissue (fascia lata tendon) which helps to strengthen the knee joint when walking and running.

Gastrocnemius

- **Position and attachments** – this is a large, superficial calf muscle with two bellies (central portion of the muscle) on the posterior of the lower leg. It is attached to the lower aspect of the posterior of the femur across the back of the knee and runs via the Achilles tendon to the calcaneum at the back of the heel.

- **Action** – plantar flexes the foot and assists in knee flexion.

Key note

The gastrocnemius muscle provides the push during fast walking and running.

Soleus

- **Position and attachments** – situated deep in the gastrocnemius in the calf, the soleus is attached to the tibia and fibula just below the back of the knee at one end and runs via the Achilles tendon to the calcaneum at the other end.

- **Action** – plantar flexes the foot.

Key note

The soleus is a thicker and flatter muscle than the gastrocnemius and accounts for the contours of the gastrocnemius being so visible.

Tibialis anterior

- **Position and attachments** – this muscle is the anterior aspect of the lower leg, attached to the outer side of the tibia at one end and the medial cuneiform and the base of the first metatarsal at the other end.

- **Action** – dorsiflexes and inverts the foot.

Key note

If the tibialis anterior muscle becomes weak, it can lead to the lower leg rolling inwards due to the collapse of the medial longitudinal arch of the foot.

Tibialis posterior

- **Position and attachments** – this muscle is found on the posterior aspect of the lower leg, very deeply situated in the calf. It is attached to the back of the tibia and fibula at one end and to the navicular, third cuneiform and second, third and fourth metatarsals at the other end.

- **Action** – assists in plantar flexion and inverts the foot.

Key note

Weakness in the tibialis posterior muscle can cause the feet to turn out from the ankles rather than the knees. This causes the muscle to stretch and the medial longitudinal arch of the foot to drop.

Peroneus longus/brevis

- **Position and attachments**– this muscle is situated on the lateral aspect of the lower leg. These muscles attach the fibula to the underneath of the first (longus) and fifth (brevis) metatarsals.

- **Action** – plantar flexes and everts the foot.

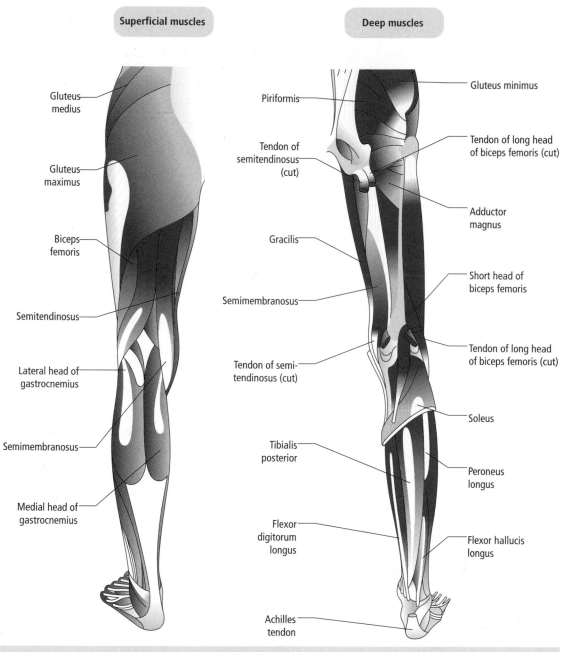

Superficial muscles

Deep muscles

Gluteus medius

Gluteus maximus

Biceps femoris

Semitendinosus

Lateral head of gastrocnemius

Semimembranosus

Medial head of gastrocnemius

Piriformis

Tendon of semitendinosus (cut)

Gracilis

Semimembranosus

Tendon of semi-tendinosus (cut)

Tibialis posterior

Flexor digitorum longus

Achilles tendon

Gluteus minimus

Tendon of long head of biceps femoris (cut)

Adductor magnus

Short head of biceps femoris

Tendon of long head of biceps femoris (cut)

Soleus

Peroneus longus

Flexor hallucis longus

Figure 4.12   Muscles of the posterior of the lower limb

Key note

Going over on to the outside of the ankle, as in a trip or a fall, can sprain the peroneal muscles in the lower leg. If the injury is not treated properly it can affect future stability of the ankle joint.

Flexor digitorum longus

- **Position and attachments** – this muscle extends from the middle third of the posterior of the tibia to the plantar surface of the second to fifth toes.
- **Action** – flexion of the toes, plantar flexion and inversion of the foot.

Flexor hallucis longus

- **Position and attachments** – this muscle extends from the distal two-thirds of the posterior fibula to the plantar surface of the big toe.
- **Action** – flexion of the big toe, plantar flexion and inversion of the foot.

Extensor digitorum longus

- **Position and attachments** – this muscle extends from the proximal two-thirds of the anterior of the fibula to the dorsal surface of the second to fifth toes.
- **Action** – extension of the second to fifth toes, dorsiflexion and eversion of the foot.

Extensor hallucis longus

- **Position and attachments** – this muscle extends from the middle third of the anterior of the fibula to the dorsal surface of the big toe.
- **Action** – extension of the big toe, dorsiflexion and inversion of the foot.

Key note

The flexor and extensor muscles of the lower leg can become weak due to excess pressure and overuse in walking and running.

The muscles of the pelvic floor

The **levator ani** and the **coccygeus** are the muscles that form the pelvic floor. These muscles support and elevate the organs of the pelvic cavity, such as the uterus and the bladder. They provide a counterbalance to increased intra-abdominal pressure which would expel the contents of the bladder, rectum and uterus.

During childbirth these muscles can become weakened and need to be strengthened by pelvic floor exercises as soon as possible after the birth.

The muscles of the anterior aspect of the trunk

Pectoralis major

■ **Position and attachments** – a thick, fan-shaped muscle covering the anterior surface of the upper chest. It attaches to the clavicle and the sternum at one end and to the humerus at the other end.

■ **Action** – adducts and medially (inwardly) rotates the arm.

Key note

Tightness in the pectoralis major muscle can cause restrictions of the chest and postural distortions such as rounded shoulders.

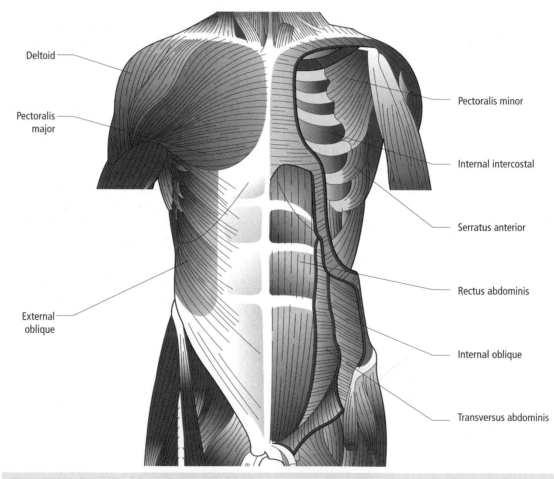

Deltoid

Pectoralis major

External oblique

Pectoralis minor

Internal intercostal

Serratus anterior

Rectus abdominis

Internal oblique

Transversus abdominis

Figure 4.13   Muscles of the anterior of the trunk

Pectoralis minor

- **Position and attachments** – this is a thin muscle that lies beneath the pectoralis major. Its fibres attach laterally and upwards from the ribs at one end to the scapula at the other end.

- **Action** – draws the shoulder downwards and forwards.

Key note

The pectoralis minor is involved in forced expiration and is therefore an accessory respiratory muscle.

Serratus anterior

- **Position and attachments** – this is a broad, curved muscle located on the side of the chest/ribcage below the axilla. It attaches the outer surface of the upper eighth or ninth rib at one end to the inner surface of the scapula, along the medial edge nearest the spine.

- **Action** – pulls the scapula downwards and forwards.

Key note

The serratus anterior has a serrated appearance which comes from attaching to separate ribs.

External oblique

- **Position and attachments** – a broad, thin sheet of muscle whose fibres slant downwards from the lower ribs to the pelvic girdle and the linea alba (tendon running from the bottom of the sternum to the pubic symphysis).

- **Action** – flexes, rotates and side-bends the trunk. It compresses the contents of the abdomen.

Key note

The external oblique muscles are often referred to as the pocket muscles as their fibres run in the direction in which you put your hands in your pockets.

Internal oblique

- **Position and attachments** – this is a broad, thin sheet of muscle located beneath the external obliques. Its fibres run up and forwards from the pelvic girdle to the lower ribs.

- **Action** – flexes, rotates and side-bends the trunk. It compresses the contents of the abdomen.

Key note

The fibres of the internal obliques are deeper and run at right angles to the external obliques.

Rectus abdominis

- **Position and attachments** – this is a long, straplike muscle that attaches to the pubic bones at one end and the ribs and the sternum at the other.
- **Action** – flexes the vertebral column, flexes the trunk (as in a sit-up) and compresses the abdominal cavity.

Key note

The rectus abdominis has three fibrous bands that give the muscle a segmented appearance and divides it into the so-called six-pack.

Transversus abdominis

- **Position and attachments** – this muscle attaches to the inner surfaces of the ribs (last six) and iliac crest at one end and extends down to the pubis via the linea alba (a long tendon that extends from the bottom of the sternum to the pubic symphysis).
- **Action** – compresses the abdominal contents and supports the organs of the abdominal cavity.

Key note

The transversus abdominis is often called the corset muscle because it wraps around the abdomen like a corset.

Diaphragm

- **Position and attachments** – this is a large, dome-shaped muscle that separates the thorax from the abdomen. It attaches to the lower part of the sternum, lower six ribs and upper three lumbar vertebrae and its fibres converge to meet on a central tendon in the abdominal cavity.
- **Action** – on contraction the diaphragm flattens to expand the volume of the thoracic cavity to assist inspiration. Upon relaxation and expiration it returns to its dome shape.

Key note

The diaphragm is unusual in that it is under both unconscious control (from the brain stem of the brain), as in the regulation of breathing, and conscious control (in that we can choose to override the brainstem control to hold our breath, sigh, sing or talk).

External intercostals

- **Position and attachments** – these are the superficial muscles that occupy and attach to the space between the ribs (called external because they are positioned on the outside).
- **Action** – help to elevate the ribcage during inhalation.

Key note

The external intercostals help to increase the depth of the thoracic cavity.

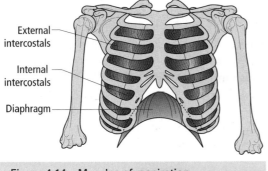

External intercostals
Internal intercostals
Diaphragm

Figure 4.14 Muscles of respiration

Internal intercostals

- **Position and attachments** – these muscles lie deep to the external intercostals (called internal because they are positioned on the inside). They occupy and attach to the spaces between the ribs.
- **Action** – depress the ribcage, which helps to move air out of the lungs when exhaling.

Key note

The external intercostals help to increase the depth of the thoracic cavity.

The muscles of the posterior aspect of the trunk

Erector spinae

- **Position and attachments** – this muscle is made up of separate bands of muscle that lie in the groove between the vertebral column and the ribs. They attach to the

sacrum and iliac crest at one end of the ribs, transverse and spinous processes of the vertebrae and the occipital bone at the other end of the ribs.

- **Action** – extension, lateral flexion and rotation of the vertebral column.

Key note

The erector spinae is a very important postural muscle as it helps to extend the spine.

Latissimus dorsi

- **Position and attachments** – this is a broad muscle that attaches to the posterior of the iliac crest and sacrum, lower six thoracic and five lumbar vertebrae at one end and the humerus at the other end.
- **Action** – extends, adducts and rotates the humerus medially.

Key note

The latissimus dorsi muscle is often referred to as the 'swimmer's muscle' as it allows extension of the arm to propel the body in water. It is also one of the major muscles of the body implicated in lower back pain due to its pelvis attachments.

Quadratus lumborum

- **Position and attachments** – this muscle attaches to the top of the posterior of the iliac crest at one end, and to the twelfth rib and transverse processes of the first four lumbar vertebrae at the other end.

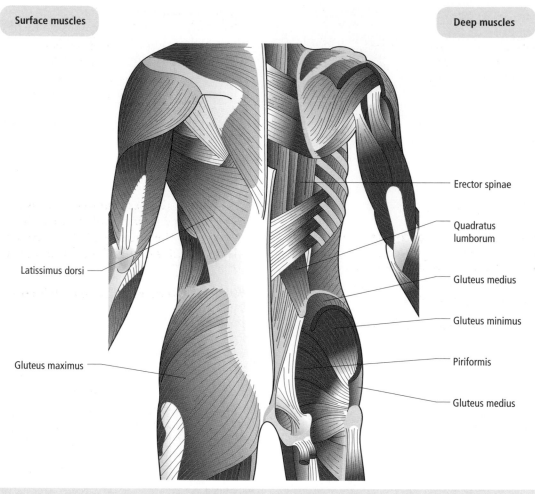

Surface muscles

Deep muscles

Erector spinae

Quadratus lumborum

Latissimus dorsi

Gluteus medius

Gluteus minimus

Piriformis

Gluteus maximus

Gluteus medius

Figure 4.15   Muscles of the posterior of the trunk

■ **Action** – lateral flexion (side-bending) of the lumbar vertebrae.

Key note

Excessive bending to the side can strain and injure the quadratus lumborum muscle.

Gluteus maximus

■ **Position and attachments** – this is a large muscle covering the buttock. It attaches to the back of the ilium along the sacroiliac joint at one end, and into the top of the femur at the other.

■ **Action** – extends the hip, abducts and laterally rotates the thigh.

Key note

The gluteus maximus is sometimes referred to as the 'speedskater's muscle' as it is powerful in extending, abducting and laterally rotating the thigh. The gluteus maximus is often implicated in postural defects such as lordosis, excess curvature in the lumbar spine.

Gluteus medius

- **Position and attachments** – this muscle is partly covered by the gluteus maximus. It attaches to the outer surface of the ilium at one end and the outer surface of the femur at the other end.

- **Action** – abducts and medially rotates the thigh.

Key note

When the gluteus medius muscle becomes tight it can create postural distortions. It pulls and depresses the pelvis towards the thigh on that side, resulting in a 'functional short leg' and a compensatory scoliosis.

Gluteus minimus

- **Position and attachments** – this muscle lies beneath the gluteus medius. Its attachments are the same as for gluteus medius: outer surface of the ilium at one end to the outer surface of the femur at the other end.

- **Action** – abducts and medially rotates the thigh.

Key note

A chronically right gluteus minimus can contribute to postural conditions such as 'functional short leg' and compensatory scoliosis, as in gluteus medius.

Piriformis

- **Position and attachments** – this is a deeply seated pelvic girdle muscle that attaches to the anterior of the sacrum at one end and the top of the femur at the other.

- **Action** – lateral rotation and abduction of the hip.

Key note

The piriformis is the largest lateral rotator of the hip. If it becomes tight, it can restrict mobility in the hip.

Psoas

- **Position and attachments** – this is a long, thick and deep pelvic muscle. It attaches the anterior transverse processes of T12–L5 (twelfth thoracic to fifth lumbar vertebrae) to the inside of the top of the femur at the other end.

- **Action** – flexes the thigh.

Iliacus

- **Position and attachments** – this is a large, fan-shaped muscle deeply situated in the pelvic girdle. It attaches to the iliac crest at one end and to the inside of the top of the femur at the other end.

- **Action** – flexes and laterally rotates the femur.

Key note

The iliacus and psoas muscles are often considered as one and may be referred to as iliopsoas. Both muscles are primary flexors of the thigh and therefore serve to advance the leg in walking.

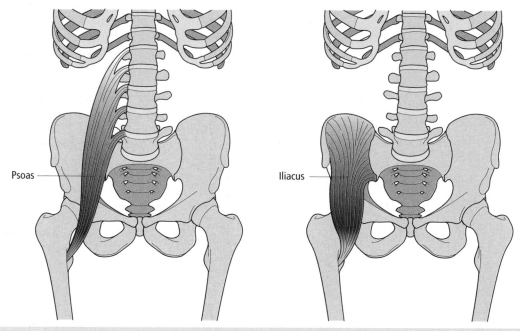

Psoas

Iliacus

Figure 4.16   Deep pelvic muscles – psoas and iliacus

Disorders of the muscular system (continued on p. 135)

(continued on p. 135)

Atony

This is a state in which the muscles are floppy and lacking their normal degree of elasticity.

Atrophy

This is the wasting of muscle tissue due to undernourishment or lack of use and diseases such as poliomyelitis.

Fibromyalgia

This is a chronic condition that produces musculoskeletal pain. Predominant symptoms include widespread musculoskeletal pain, lethargy and fatigue. Other characteristic features include a non-refreshing sleep pattern in which the patient feels exhausted and more tired than later in the day and interrupted sleep.

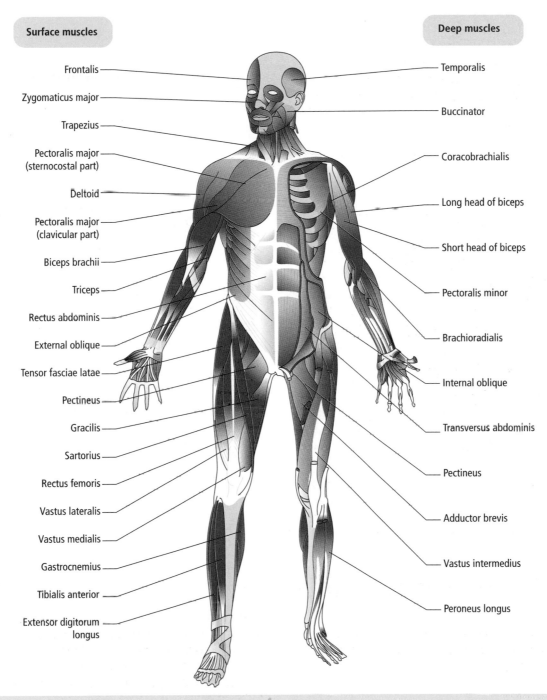

Surface muscles

Frontalis
Zygomaticus major
Trapezius
Pectoralis major (sternocostal part)
Deltoid
Pectoralis major (clavicular part)
Biceps brachii
Triceps
Rectus abdominis
External oblique
Tensor fasciae latae
Pectineus
Gracilis
Sartorius
Rectus femoris
Vastus lateralis
Vastus medialis
Gastrocnemius
Tibialis anterior
Extensor digitorum longus

Deep muscles

Temporalis
Buccinator
Coracobrachialis
Long head of biceps
Short head of biceps
Pectoralis minor
Brachioradialis
Internal oblique
Transversus abdominis
Pectineus
Adductor brevis
Vastus intermedius
Peroneus longus

Figure 4.17   Anterior muscles of the body

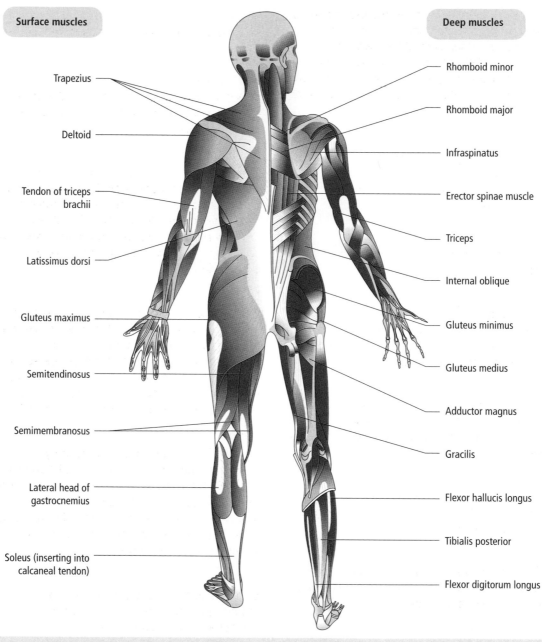

Surface muscles

Trapezius

Deltoid

Tendon of triceps brachii

Latissimus dorsi

Gluteus maximus

Semitendinosus

Semimembranosus

Lateral head of gastrocnemius

Soleus (inserting into calcaneal tendon)

Deep muscles

Rhomboid minor

Rhomboid major

Infraspinatus

Erector spinae muscle

Triceps

Internal oblique

Gluteus minimus

Gluteus medius

Adductor magnus

Gracilis

Flexor hallucis longus

Tibialis posterior

Flexor digitorum longus

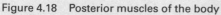

Figure 4.18   Posterior muscles of the body

■ Other recognised symptoms include early morning stiffness, pins and needles sensation, unexplained headaches, poor concentration, memory loss, low mood, urinary frequency, abdominal pain, irritable bowel syndrome. Anxiety and depression are also common.

Fibrositis

Fibrositis is an inflammatory condition of the fibrous connective tissues, especially in the muscle fascia (also known as muscular rheumatism).

Myositis

This condition is the inflammation of a skeletal muscle.

Muscle cramp

This is an acute, painful contraction of a single muscle or group of muscles. Cramp is often associated with a mineral deficiency, an irritated nerve or muscle fatigue.

Muscle fatigue

This is the loss of the ability of a muscle to contract efficiently due to insufficient oxygen, exhaustion of energy supply and the accumulation of lactic acid.

Muscle spasm

This is an increase in muscle tension due to excessive motor nerve activity, resulting in a knot in the muscle.

Muscular atrophy

This is the wasting away of muscles due to poor nutrition, lack of use or a dysfunction of the motor nerve impulses.

Muscular dystrophy

This is a progressively crippling disease in which the muscles gradually weaken and atrophy. The cause is unknown.

Rupture

A rupture is the tearing of a muscle fascia or tendon.

Shin splints

This is a soreness in the front of the lower leg due to straining of the flexor muscles used in walking.

Spasticity

This is characterised by an increase in muscle tone and stiffness. In severe cases movements may become uncoordinated and involve a nervous dysfunction. Spasticity involves muscles with excessive tone and is a condition often associated with nervous dysfunction.

Sprain

This is a complete or incomplete tear in the ligaments around a joint. It usually follows a sudden, sharp twist to the joint that stretches the ligaments and ruptures some or all of its fibres. Sprains commonly occur in the ankle, wrist and back, where there is localised pain, swelling and loss of mobility.

Strain

A strain is an injury that is caused by excessive stretching or working of a muscle or tendon that results in a partial or complete tear. Symptoms include pain, swelling, tenderness and stiffness in the affected area. Muscle strains are more common in the lower back and the neck.

Stress

Stress is excessive muscular tension resulting in tight, painful muscles and restricted joint movements.

Tendinitis

This is the inflammation of a tendon, accompanied by pain and swelling.

Tennis elbow

This condition is the inflammation of the tendons that attach the extensor muscles of the forearm at the elbow joint.

Torticollis

This is a condition in which the neck muscles (sternomastoids) contract involuntarily. It is commonly called wryneck.

Interrelationships with other systems

The muscular system is linked to the following body systems:

Cells and tissues/histology

There are three types of muscle tissue in the body – skeletal or voluntary muscle, smooth and cardiac. Fasciae, tendons and ligaments are all made from connective tissue and serve a function in muscle attachments.

Skeletal system

Bones and joints provide the leverage in a movement and the muscles provide the pull on the bone to effect the movement.

Circulatory system

The circulatory system is responsible for delivering oxygen, glycogen and water to the working muscles. It also transports waste products, such as carbon dioxide and lactic acid, away from the muscles.

Respiratory system

The respiratory system provides the working muscles with vital oxygen which is transported in the blood to be combined with glycogen to release energy.

Nervous system

Muscles rely on nervous stimulation in order to function. Skeletal muscles are moved as a result of nervous stimulus which they receive from the brain via a motor nerve.

Digestive system

The energy needed for muscle contraction is obtained principally from carbohydrate digestion. Carbohydrates are broken down and glucose, which is not required immediately by the body, is converted into glycogen and stored in the liver and muscle.

Key words associated with the muscular system

skeletal/voluntary muscle tissue	sarcomere	origin
cardiac muscle tissue	motor nerve	insertion
smooth/involuntary muscle	motor point	antagonists
fasciculi	neurotransmitter	agonist/prime mover
epimysium	glycogen	synergists
endomysium	lactic acid	fixators
perimysium	muscle fatigue	isometric contraction
myofibrils	muscle tone	isotonic contraction
actin	tendon	concentric contractions
myosin	ligament	eccentric contractions
	fascia	

Summary of the muscular system

- The muscular system is comprised mainly of **skeletal** or **voluntary muscle tissue** that is primarily attached to bones.

- The other types of muscle tissue are **cardiac muscle tissue**, found in the wall of the heart, and **smooth muscle tissue**, located in the wall of the stomach and small intestines.

- Through contraction muscle performs three important functions – movement, maintaining posture and heat production.

- **Voluntary** or **skeletal muscle tissue** consists of muscle fibres held together by fibrous connective tissue and penetrated by numerous tiny blood vessels and nerves.

- Voluntary muscle tissue is made up of bands of elastic or contractile tissue bound together in bundles and enclosed by a connective tissue sheath.

- Each muscle fibre is enclosed in an individual wrapping of connective tissue called the **endomysium**.

- The muscle fibres are wrapped together in bundles, known as **fasciculi**, and are covered by the **perimysium** (fibrous sheath) and then gathered to form the muscle belly (main part of the muscle) with its own sheath, the fascia **epimysium**.

- Each skeletal muscle fibre is made up of thin fibres called **myofibrils** which consist of two different types of protein strands called **actin** and **myosin**. This gives skeletal muscle its striated or striped appearance.

- Muscle fibre contraction results from a sliding movement within the **myofibrils** in which **actin** and **myosin** filaments merge.

- Skeletal muscle is moved as a result of nervous stimulus received from the brain via a motor nerve.

- Each nerve fibre ends in a **motor point**, which is the part through which the stimulus is given to contract.

- The muscle cells in smooth or involuntary muscle are spindle-shaped and tapered at both ends, with each muscle cell containing one centrally located, oval-shaped nucleus.

- **Smooth muscle** contracts or relaxes in response to nerve impulses, stretching or hormones.

- **Cardiac muscle** is found only in the heart and, like skeletal muscle, it is striated. However, it is branched in structure and has intercalated discs in between each muscle cell.

- The contraction of cardiac muscle is regulated by nerves and hormones.

- During muscular contraction a sliding movement occurs within the contractile fibres (**myofibrils**).

- The **actin** filaments move in towards the **myosin** and cause the muscle fibres to shorten and thicken.

- During relaxation the muscle fibres elongate and return to their original shape.

- The energy needed for muscle contraction comes from **glycogen** (stored in the liver and the muscles) and oxygen.

- If insufficient oxygen is available to a working muscle a waste product called **lactic acid** forms which can cause a muscle to ache.

- The term **muscle fatigue** is defined as the loss of ability of a muscle to contract efficiently due to insufficient oxygen, exhaustion of glucose and the accumulation of lactic acid.

- During exercise the circulatory and respiratory systems adjust to cope with the increased oxygen demands of the body. More blood is distributed to the working muscles and the rate and depth of breathing is increased.

- When muscle tissue is warm, muscle contraction will occur faster due to the increase in circulation and acceleration of chemical reactions.

- Conversely, when muscle tissue is cooled, the chemical reactions and circulation slow down.

- The term **muscle tone** is the state of partial contraction of a muscle to help maintain body posture.

- Good muscle tone can be recognised by the muscles appearing firm and rounded.

- Poor muscle tone may be recognised by the muscles appearing loose and flattened.

- **Tendons** are tough bands of white fibrous tissue that link muscle to bone.

- Unlike muscle they are inelastic and therefore do not stretch.

- **Ligaments** are strong, fibrous, elastic tissues that link bones together and therefore stabilise joints.

- **Fascia** consists of fibrous connective tissue that envelops a muscle and provides a pathway for nerves, blood vessels and lymphatic vessels.

- **Fascia**, therefore, plays a key role in maintaining the health of a muscle.

- Muscle attachments are known as **origin** and **insertion**.

- The **origin** is the end of the muscle closest to the centre of the body and the **insertion** is the furthest attachment.

- The **insertion** is generally the most movable point and the point at which the muscle work is done.

- In movement coordination muscles work in pairs or groups.

- Muscles are classified by functions as **agonists** (prime movers), **antagonists**, **synergists** and **fixators** (stabilisers).

- **Antagonists** are two muscles or sets of muscles pulling in opposite directions to each other, with one relaxing to allow the other to contract.

- **Agonist/prime mover** is known as the main activating muscle.

- **Synergist** refers to muscles on the same side of a joint that work together to perform the same movements.

- Muscular contractions can be **isometric** or **isotonic**.

- **Isometric contraction** is when the muscle works without actual movements (postural muscles).

- **Isotonic contraction** is when the muscle's force is considered to be constant but the muscle length changes.

- There are two types of isotonic contraction – **concentric contractions** (towards the centre) and **eccentric contractions** (away from the centre).

Multiple-choice questions

The muscular system

1. **Which of the following is not a function of the muscular system?**

   a   movement
   b   exchanging of gases
   c   production of heat
   d   maintaining posture

2. **A voluntary muscle will only contract if a stimulus is applied to it via a:**

   a   sensory nerve
   b   motor nerve
   c   mixed nerve
   d   none of the above

3. **A tendon attaches**

   a   muscle to bone
   b   muscle to ligament
   c   bone to bone
   d   none of the above

4. **Where would you not find involuntary muscle tissue?**

   a   stomach
   b   bladder
   c   brain
   d   heart

5. **The fuel for muscle contraction is provided by:**

   a   ATP
   b   glucose
   c   pyruvic acid
   d   actin and myosin

6. **Which of the following statements is true?**

   a   Upon voluntary muscle contraction muscle fibres elongate
   b   The attachment of myosin to actin requires the mineral sodium
   c   The merging of actin and myosin filaments causes the muscle fibres to shorten and thicken on contraction
   d   The force of muscle contraction depends on where the muscle fibres are located

7. **The condition muscle fatigue is caused by:**

   a   insufficient oxygen
   b   exhaustion of energy supply
   c   accumulation of lactic acid
   d   all of the above

8. **The state of continuous partial contraction of muscles is known as:**

   a   atrophy
   b   muscle tone
   c   hypertrophy
   d   none of the above

9. **Which of the following statements is false?**

   a   During exercise there is an increased return of venous blood to the heart
   b   During exercise a muscle may receive as much as 15 times its normal flow of blood
   c   The presence of lactic acid in the blood stimulates the respiratory centre in the brain, decreasing the rate and depth of breath
   d   The rate and depth of breath remains above normal for a while after strenuous exercise has ceased

10. **Which of the following statements is true?**

   a   muscles with less than the normal degree of tone are said to be spastic
   b   good muscle tone may be recognised by the muscles appearing firm and rounded
   c   an increase in the size and diameter of the muscle fibres leads to a condition called atrophy
   d   poor muscle tone is a cause of muscle cramps

Muscles of the face

11. The name of the muscle that closes the mouth is:

a   risorius
b   orbicularis oris
c   orbicularis oculi
d   levator anguli oris

12. The name of the muscle that turns the head to the opposite side is:

a   occipitalis
b   platysma
c   sternocleidomastoid
d   frontalis

13. The action of the zygomaticus muscle is to:

a   pull the corner of the mouth sideways and upwards
b   draw the angle of the mouth upwards
c   draw the corner of the mouth downwards
d   raise the jaw

14. The facial expression associated with the mentalis muscle is:

a   pouting
b   smiling
c   laughing
d   grinning

15. The facial expression associated with the corrugator muscle is:

a   puzzled
b   frowning
c   snarling
d   sadness

16. The position of the buccinator muscle is:

a   under the corners of the mouth
b   at the sides of the nose
c   over the centre of the chin
d   in the cheek

17. Which of the following muscles does not have an action on the mouth?

a   levator labii superioris
b   depressor anguli oris
c   levator anguli oris
d   procerus

18. The main muscle of mastication in the face is:

a   buccinator
b   mentalis
c   masseter
d   none of the above

Muscles of the body

19. The name of the large, triangular-shaped muscle in the upper back that raises the shoulder girdle is:

a   splenius cervicis
b   trapezius
c   levator scapula
d   splenius capitis

20. The action of the rhomboid muscles is to:

a   elevate and adduct the scapula
b   adduct the humerus
c   abduct the scapula
d   abduct the humerus

21. The muscle located in the depression above the spine of the scapula is:

a   teres major
b   supraspinatus
c   teres minor
d   infraspinatus

22. Which of the following muscles does not flex the forearm?

a   biceps
b   brachialis
c   pronator teres
d   triceps

23. The action of the hamstring muscles is:

a flexion of the hip and knee
b flexion of the knee and extension of the hip
c extension of the knee and flexion of the hip
d adduction of the thigh and flexion of the hip

24. The muscle that runs down the lateral aspect of the thigh and is attached to a broad sheet of tendon that strengthens the knee is:

a gracilis
b tensor fascia lata
c sartorius
d piriformis

25. Which of the following muscles is not a plantar flexor of the foot?

a gastrocnemius
b tibialis anterior
c peroneus longus
d soleus

26. The deepest of the abdominal muscles is the:

a transverse abdominis
b rectus abdominis
c external obliques
d internal obliques

27. The important postural muscle of the back that extends the spine is:

a latissimus dorsi
b psoas
c iliacus
d erector spinae

28. Which of the following muscles attached to the clavicle and sternum at one end and to the humerus at the other end?

a serratus anterior
b pectoralis major
c pectoralis minor
d platysma

29. Which of the following muscles is not involved in respiration?

a diaphragm
b internal obliques
c internal intercostals
d external intercostals

30. The action of the gluteus maximus muscle is:

a abduction and medial rotation of the thigh
b extension of the hip and lateral rotation of the thigh
c flexion and lateral rotation of the femur
d lateral flexion of the spine

31. The only muscles that move the head but does not attach to any vertebrae is:

a splenius capitis
b splenius cervicis
c trapezius
d sternocleidomastoid

32. Which of the following muscles does not rotate the humerus?

a supraspinatus
b subscapularis
c infraspinatus
d teres minor

33. If one shoulder is elevated which muscle would you most likely find contracted?

a pectoralis major
b serratus anterior
c levator scapula
d supraspinatus

34. Which of the following muscles extends the big toe?

a extensor digitorum longus
b flexor digitorum longus
c extensor hallicus longus
d flexor hallicus longus

the cardiovascular system

Introduction

The cardiovascular system is the body's transport system and comprises blood, blood vessels and the heart. Blood provides the fluid environment for our body's cells and is transported in specialised tubes called blood vessels. The heart acts like a pump which keeps the blood circulating around the body in a constant circuit.

Objectives

By the end of this chapter, you will be able to recall and understand the following knowledge:

- the composition and functions of blood
- the structural and functional significance of the different types of blood cells
- the structural and functional differences between the different blood vessels
- major blood vessels of the heart
- the pulmonary and systemic blood circulation
- blood pressure and the pulse rate
- the interrelationships between the cardiovascular and other body systems
- circulatory disorders.

Blood

Blood is the fluid tissue or medium in which all materials are transported to and from individual cells in the body. Blood is, therefore, the chief transport system of the body.

The percentage composition of blood

Blood is composed of 55 per cent of fluid or plasma, which is a clear, pale-yellow, slightly alkaline fluid consisting of the following substances:

- 91 per cent of plasma is water
- 9 per cent remaining consists of dissolved blood proteins, waste, digested food materials, mineral salts and hormones
- 45 per cent of blood is made up of the blood cells erythrocytes, leucocytes and thrombocytes.

Functions of blood

There are four main functions of blood:

- transport
- defence
- regulation
- clotting.

Transport

Blood is the primary transport medium for a variety of substances that travel throughout the body.

- Oxygen is carried from the lungs to the cells of the body in red blood cells.
- Carbon dioxide is carried from the body's cells to the lungs.

- Nutrients such as glucose, amino acids, vitamins and minerals are carried from the small intestine to the cells of the body.
- Cellular waste such as water, carbon dioxide, lactic acid and urea are carried in the blood to be excreted.
- Hormones, which are internal secretions that help to control important body processes, are transported by the blood to target organs.

Key note

Red blood cells are called erythrocytes and they contain the red protein pigment haemoglobin that combines with oxygen to form oxyhaemoglobin. The pigment haemoglobin assists the function of the erythrocyte in transporting oxygen from the lungs to the body's cells and carrying carbon dioxide away.

Defence

White blood cells are collectively called leucocytes and they play a major role in combating disease and fighting infection.

Key note

White blood cells are known as phagocytes as they have the ability to engulf and ingest microorganisms which invade the body and cause disease. Specialised white blood cells called lymphocytes produce antibodies to protect the body against infection.

Regulation

Blood helps to regulate heat in the body by absorbing large quantities of heat produced by the liver and the muscles. This is then transported around the body to help to maintain a constant internal temperature. Blood also helps to regulate the body's pH balance.

Clotting

Clotting is an effective mechanism in controlling blood loss from blood vessels when they have become damaged as in a cut. Specialised blood cells called thrombocytes, or platelets, form a clot around the damaged area to prevent the body from losing too much blood and to prevent the entry of bacteria.

Blood cells

There are three types of blood cells:

- **erythrocytes** – red blood cells
- **leucocytes** – white blood cells
- **thrombocytes** – platelets.

Erythrocytes

Erythrocytes are disc-shaped structures and make up more than 90 per cent of the formed elements in blood. They are formed in red bone marrow and contain the iron-protein compound haemoglobin.

Old and worn-out erythrocytes are destroyed in the liver and the spleen. The haemoglobin is broken down and the iron within it is retained for further haemoglobin synthesis. Erythrocytes have a life span of only about four months and, therefore, have to be continually replaced. The function of erythrocytes is to transport the gases of respiration (they transport oxygen to the cells and carry carbon dioxide away from the cells).

Figure 5.1 An erythrocyte

Leucocytes

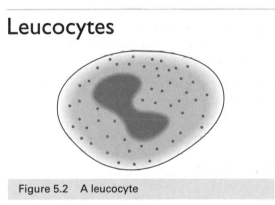

Figure 5.2 A leucocyte

Leucocytes are the largest of all the blood cells and appear white due to their lack of haemoglobin. They have a nucleus and are generally more numerous than erythrocytes.

There are two main categories of leucocytes:

- **Granulocytes** – these account for about 75 per cent of white blood cells and can be further divided into **neutrophils**, **eosinophils** and **basophils**.

- **Agranulocytes** – these can be divided into **lymphocytes**, which account for about 20 per cent of all white blood cells, and **monocytes**, which account for about 5 per cent of white blood cells.

Leucocytes usually only survive for a few hours, but in a healthy body some can live for months or even years. The main function of leucocytes is to protect the body against infection and disease in a process known as phagocytosis, which means to engulf and ingest microbes, dead cells and tissue.

Thrombocytes

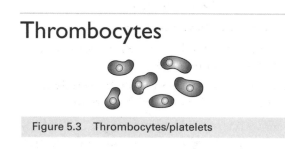

Figure 5.3   Thrombocytes/platelets

Thrombocytes are also known as **platelets**. These are small fragments of cells and are the smallest cellular elements of the blood. They are formed in bone marrow and are disc-shaped, with no nucleus. Thrombocytes normally have a short life span of just five to nine days.

They are very significant in the blood-clotting process as they initiate the chemical reaction that leads to the formation of a blood clot. Platelets stop the loss of blood from a damaged blood vessel in the following way:

- Platelets gather where a blood vessel is injured and red cells are flowing out.

- The first platelets to arrive form a plug across the opening and release chemicals that convert fibrinogen (a coagulation factor) to fibrin.

- Fibrin forms a mesh of needlelike fibres that trap platelets and other blood cells, creating an insoluble clot.

Blood vessels

Blood flows round the body by the pumping action of the heart and is carried in vessels known as arteries, veins and capillaries.

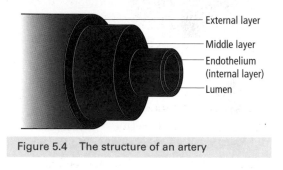

Figure 5.4   The structure of an artery

Arteries

- Arteries carry blood away from the heart.
- Blood is carried under high pressure.
- Arteries have thick, muscular and elastic walls to withstand pressure.
- Arteries have no valves, except at the base of the pulmonary artery where they leave the heart.
- Arteries (except the pulmonary artery) carry oxygenated blood to the lungs.
- Arteries are generally deep-seated, except where they cross over a pulse spot.
- Arteries give rise to small blood vessels called arterioles which deliver blood to the capillaries.

Veins

- Veins carry blood towards the heart.
- Blood is carried under low pressure.
- Veins have thinner muscular walls.
- Veins have valves at intervals to prevent the backflow of blood.

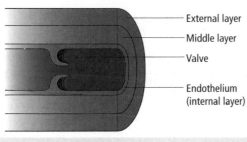

Figure 5.5   The structure of a vein

- Veins (except the pulmonary veins) carry deoxygenated blood from the lungs.
- Veins are generally superficial, not deep-seated.
- Veins form finer blood vessels called venules which continue from capillaries.

Key note

Both arteries and veins have three layers (external, middle and internal), but because an artery must contain the pressure of blood pumped from the heart, its walls are thicker and more elastic.

Capillaries

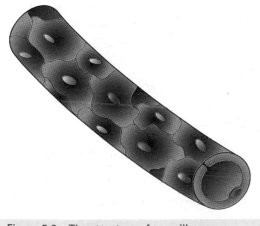

Figure 5.6   The structure of a capillary

- Capillaries are the smallest vessels.
- Capillaries unite arterioles and venules, forming a network in the tissues.
- The wall of a capillary vessel is only a single layer of cells thick. It is, therefore, sufficiently thin to allow the process of diffusion of dissolved substances to and from the tissues to occur.
- Capillaries have no valves.
- Blood is carried under low pressure but higher than in veins.
- Capillaries are responsible for supplying the cells and tissues with nutrients.

Oxygenated blood flowing through the arteries appears bright red in colour due to the oxygen pigment haemoglobin. As it moves through the capillaries, it offloads some of its oxygen and picks up carbon dioxide. This explains why blood flow in veins appears darker.

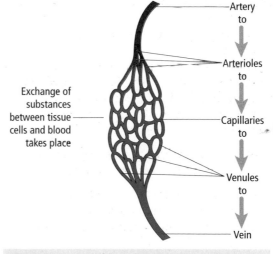

Figure 5.7   Blood flow from an artery to a vein

Key note

The key function of a capillary is to permit the exchange of nutrients and waste between the blood and tissue cells. Substances such as oxygen, vitamins, minerals and amino acids pass through to the tissue fluid to nourish the nearby cells, and substances such as carbon dioxide and waste are passed out of the cells. This exchange of nutrients can only occur through the semipermeable membrane of a capillary as the walls of arteries and veins are too thick.

The heart

The heart is a hollow organ made up of cardiac muscle tissue which lies in the thorax above the diaphragm and between the lungs.

Composition of the heart

The heart is composed of three layers of tissue:

Pericardium: the outer layer

This is a double-layered bag enclosing a cavity filled with pericardial fluid which reduces friction as the heart moves during its beating.

Myocardium: the middle layer

This is a strong layer of cardiac muscle which makes up the bulk of the heart.

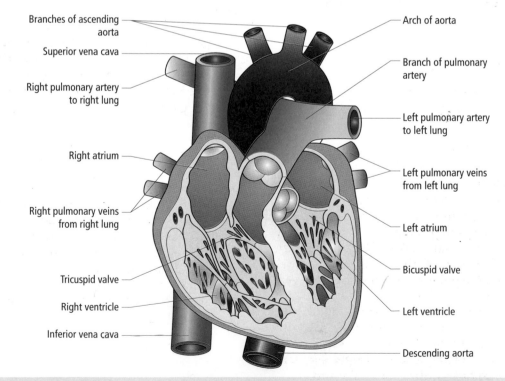

Branches of ascending aorta
Superior vena cava
Right pulmonary artery to right lung
Right atrium
Right pulmonary veins from right lung
Tricuspid valve
Right ventricle
Inferior vena cava

Arch of aorta
Branch of pulmonary artery
Left pulmonary artery to left lung
Left pulmonary veins from left lung
Left atrium
Bicuspid valve
Left ventricle
Descending aorta

Figure 5.8   The structure of the heart

Endocardium: the inner layer

This lines the heart's cavities and is continuous with the lining of the blood vessels. The heart is divided into a right and left side by a partition called a septum and each side is further divided into a thin-walled atrium above and a thick-walled ventricle below. The top chambers of the heart (the atria) take in blood from the body from the large veins and pump it to the bottom chambers. The lower chambers, the ventricles, pump blood to the body's organs and tissues. There are four sets of valves that regulate the flow of blood through the heart:

■ The **tricuspid valve** is found between the right atrium and the right ventricle.

■ The **bicuspid** or **mitral valve** is found between the left atrium and the left ventricle.

■ The **aortic valve** is found between the left ventricle and the aorta.

■ The **pulmonary valve** is found between the pulmonary artery and the right ventricle.

The bicuspid and tricuspid valves (also known as the atrioventricular valves) help to maintain the direction of blood flow through the heart by allowing blood to flow into the ventricles but keeping it from returning to the atria.

The aortic and pulmonary valves are known as the semi-lunar valves. They control the blood flow out of the ventricles into the aorta and the pulmonary arteries and prevent any backflow of blood into the ventricles. These valves open in response to pressure generated when the blood leaves the ventricles. The heart muscle is supplied by the two coronary arteries (right and left) which originate at the base of the aorta.

Key note

If either of the coronary arteries is unable to supply sufficient blood to the heart muscle a heart attack occurs. The most common site of a heart attack is the anterior or inferior part of the left ventricle.

Blood flow through the heart

Blood moves into and out of the heart in a well-coordinated and precisely timed rhythm.

For descriptive purposes the blood flow can be divided into three stages:

Stage 1

Deoxygenated blood from the body enters the superior and inferior vena cava and flows into the right atrium. When the right atrium is full, it empties through the tricuspid valve into the right ventricle.

Stage 2

When the right ventricle is full, it contracts and pushes blood through the pulmonary valve into the pulmonary artery. The pulmonary artery divides into the right and left branches and takes blood to both lungs where the blood becomes oxygenated. The four pulmonary veins leave the lungs, carrying oxygen-rich blood back to the left atrium.

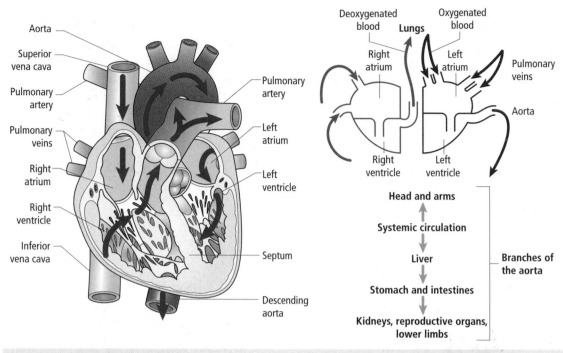

Figure 5.9   Blood flow through the heart

Stage 3

This process takes place at the same time as the process described in stage 1. Oxygen-rich blood leaves the left atrium and passes through the left ventricle via the bicuspid or mitral valve. When the left ventricle is full it contracts, forcing blood through the aortic valve into the aorta and to all parts of the body (except the lungs). The walls of the left ventricle are thicker in order to provide the extra strength to push blood out of the heart and around the body.

Function of the heart

The function of the heart is to maintain a constant circulation of blood throughout the body. The heart acts as a pump and its action consists of a series of events known as the cardiac cycle.

The cardiac cycle

The cardiac cycle is the sequence of events between one heartbeat and the next and is normally less than a second in duration.

■ During a cardiac cycle the atria contract simultaneously and force blood into the relaxed ventricles.

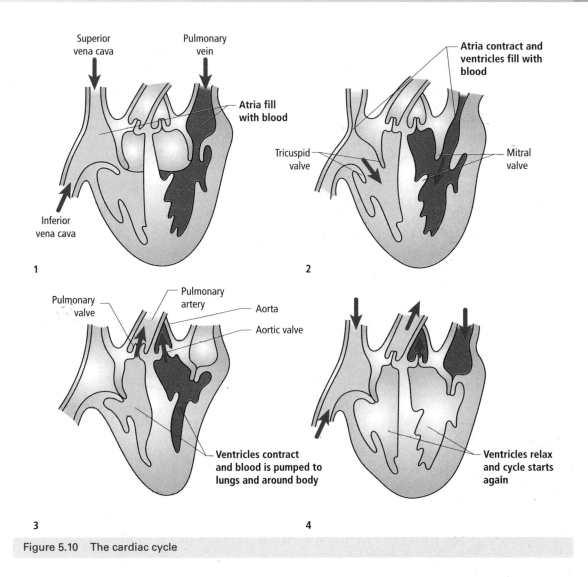

Figure 5.10 The cardiac cycle

- The ventricles then contract very strongly and pump blood out through the aorta and the pulmonary artery.

- During ventricular contraction the atria relax and fill up again with blood.

The heart rate can be determined by the number of cardiac cycles per minute. In an average healthy person this is likely to be between 60 and 70 cycles or beats per minute.

The heart has its own built-in rhythm. The coordinated rhythm of the heart is initiated by the built-in electrical system in the sinoatrial (SA) node which sets the pace of the heart rate. The signal originates in the right atrium and travels to the left atrium, causing the atria to contract. At the precise moment the atria have completed their contraction, the signal travels the atrioventricular (AV) bundle to the right ventricle and into the left ventricle, causing the ventricles to contract.

Key note

If difficulty develops within the electrical system of the SA node, a device known as a pacemaker can be implanted to assist or take over initiation of the signal.

Heart sounds

Heart sounds may be heard through a stethoscope. Closure of the heart valves produces two main sounds:

- The first is a low-pitched 'lubb' which is generated by the closing of the bicuspid and tricuspid valves.
- The second is a higher-pitched 'dubb' caused by the closing of the aortic and pulmonary valves.

Blood is transported as part of a double circuit and consists of two separate systems which are joined only at the heart.

The pulmonary circulation

The pulmonary circulation is the circulatory system's contribution to respiration. This consists of the circulation of deoxygenated blood from the right ventricle of the heart to the lungs via the pulmonary arteries. It becomes oxygenated here and is then returned to the left atrium by the pulmonary veins to be passed to the aorta for the general or systemic circulation.

The pulmonary circulation is essentially the circulatory system between the heart and the lungs where a high concentration of blood oxygen is restored and the concentration of carbon dioxide in the blood is lowered.

The general or systemic circulation

The systemic circuit is the largest circulatory system and carries oxygenated blood from the left ventricle of the heart through to the aorta. Oxygenated blood is then passed around the body through the various branches of the aorta. Deoxygenated blood is returned to the right atrium via the superior and inferior vena cava to be passed to the right ventricle to enter the pulmonary circuit. The function of the systemic circulation is to bring nutrients and oxygen to all systems of the body and carry waste materials away from the tissues for elimination.

Key note

The increase in blood flow during a massage can help to bring fresh oxygen and nutrients into the tissues via the arterial circulation and aid the removal of waste products via the venous circulation. Blood circulation can, therefore, help to improve the condition of the skin and, combined with massage, can also help to improve the muscle tone.

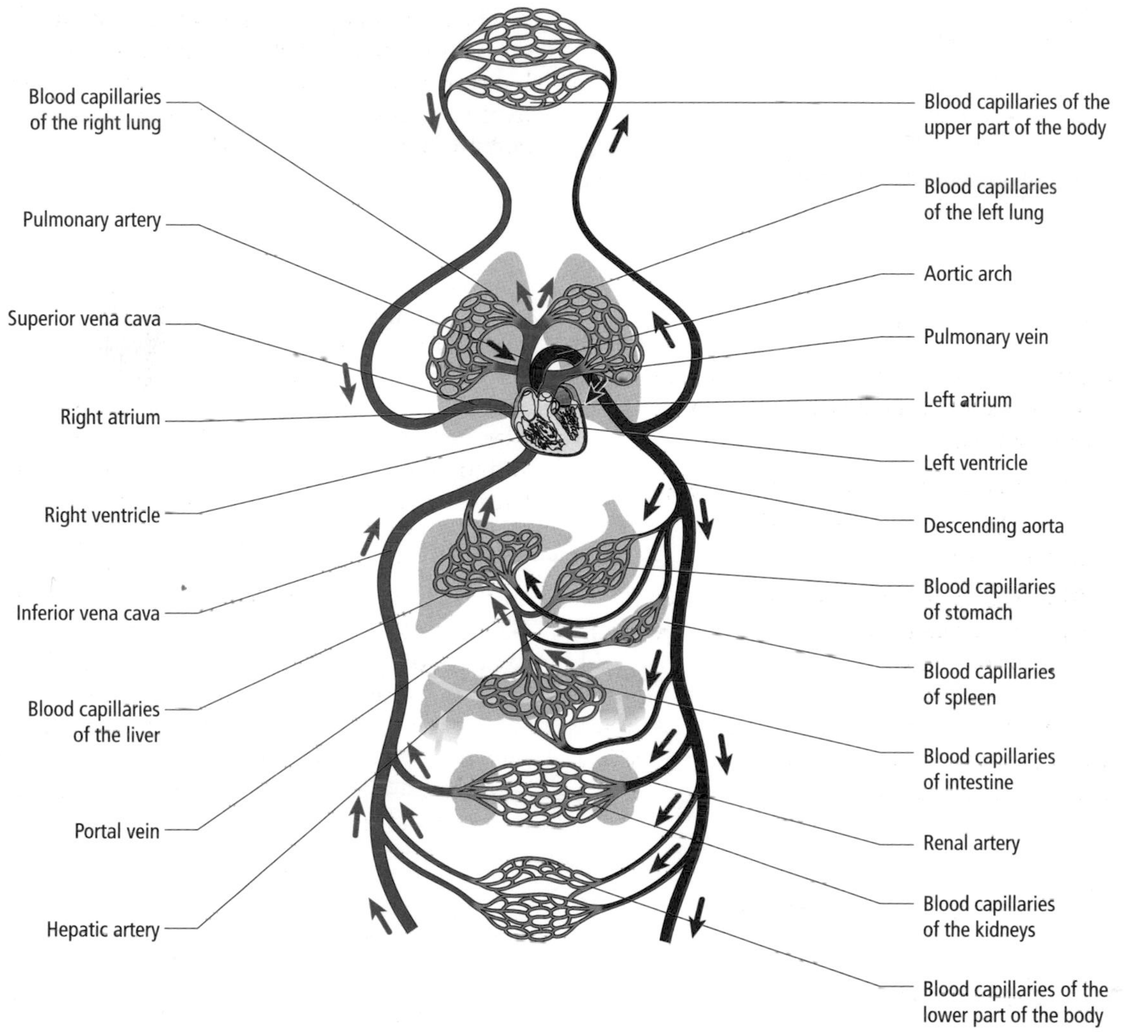
Blood capillaries of the right lung
Pulmonary artery
Superior vena cava
Right atrium
Right ventricle
Inferior vena cava
Blood capillaries of the liver
Portal vein
Hepatic artery
Blood capillaries of the upper part of the body
Blood capillaries of the left lung
Aortic arch
Pulmonary vein
Left atrium
Left ventricle
Descending aorta
Blood capillaries of stomach
Blood capillaries of spleen
Blood capillaries of intestine
Renal artery
Blood capillaries of the kidneys
Blood capillaries of the lower part of the body

Figure 5.11 The pulmonary and systemic circulation

The portal circulation

Located within the systemic circuit is the portal circulation which collects blood from the digestive organs (stomach, intestines, gall bladder, pancreas and spleen) and delivers this blood to the liver for processing, via the hepatic portal veins. As the liver has a key function in maintaining proper concentrations of glucose, fat and protein in the blood, the hepatic portal system allows the blood from the digestive organs to take a detour through the liver to process these substances before they enter the systemic circulation.

Main arteries

The aorta is the main artery of the systemic circuit which carries oxygenated blood around the body. It is divided into three main branches which subdivide into branches to supply the whole body:

- The ascending part has branches which supply the head, neck and the top of the arms.

- The descending thoracic part of the aorta has branches which supply organs of the thorax.

- The descending abdominal part has branches which supply the legs and organs of the digestive, renal and reproductive systems.

The names of most major arteries are derived from the anatomical structures they serve, such as the femoral artery which is found close

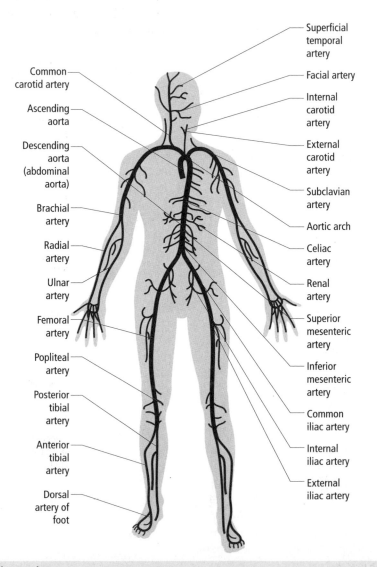

Common carotid artery

Ascending aorta

Descending aorta (abdominal aorta)

Brachial artery

Radial artery

Ulnar artery

Femoral artery

Popliteal artery

Posterior tibial artery

Anterior tibial artery

Dorsal artery of foot

Superficial temporal artery

Facial artery

Internal carotid artery

External carotid artery

Subclavian artery

Aortic arch

Celiac artery

Renal artery

Superior mesenteric artery

Inferior mesenteric artery

Common iliac artery

Internal iliac artery

External iliac artery

Figure 5.12   Main arteries

to the femur. Arteries generally lie deeply seated. They are found on both sides of the body and are identified as either right or left.

Main veins

The major veins of the body are the **superior** and **inferior vena cava** which convey deoxygenated blood from the other veins to the right atrium of the heart.

■ The **inferior vena cava** is formed by the joining of the right and left **iliac veins**. It receives blood from the lower parts of the body below the diaphragm.

■ The **superior vena cava** originates at the junction of the two **innominate (brachiocephalic)** veins. It drains blood from the upper parts of the body (head, neck, thorax and arms).

Like arteries, veins are also named for their locations and have two branches (right and left). Veins are more superficially placed than arteries.

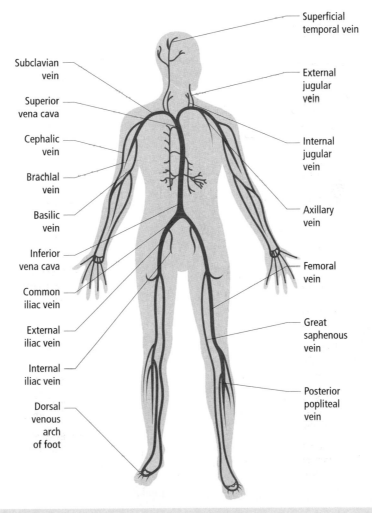

Subclavian vein
Superior vena cava
Cephalic vein
Brachlal vein
Basilic vein
Inferior vena cava
Common iliac vein
External iliac vein
Internal iliac vein
Dorsal venous arch of foot

Superficial temporal vein
External jugular vein
Internal jugular vein
Axillary vein
Femoral vein
Great saphenous vein
Posterior popliteal vein

Figure 5.13   Main veins

Blood vessels of the head and neck

Blood is supplied to parts within the neck, head and brain through branches of the **subclavian** and **common carotid arteries**. The **common carotid artery** extends from the **brachiocephalic artery**. It extends on each side of the neck and divides at the level of the larynx into two branches:

- The **internal carotid artery** passes through the temporal bone of the skull to supply oxygenated blood to the brain, eyes, forehead and part of the nose.

- The **external carotid artery** is divided into branches (facial, temporal and occipital arteries) which supply the skin and muscles of the face, side and back of the head respectively. This vessel also supplies more superficial parts and structures of the head and neck. These include the salivary glands, scalp, teeth, nose, throat, tongue and thyroid gland.

The **vertebral arteries** are a main division of the **subclavian artery**. They arise from the subclavian arteries in the base of the neck, near the tip of the lungs, and pass upwards through the openings (foramina) of transverse processes of the cervical vertebrae where they unite to form a single **basilar artery**. The **basilar artery** then terminates by dividing into two posterior **cerebral arteries** that supply the occipital and temporal lobes of the cerebrum.

The majority of blood draining from the head is passed into three pairs of veins:

- **external jugular veins**

- **internal jugular veins**

- **vertebral veins**.

Within the brain all veins lead to the internal jugular veins.

The **external jugular veins** are smaller than the **internal jugular veins** and lie superficial to them. They receive blood from superficial regions of the face, scalp and neck. The **external jugular veins** descend on either side of the neck, passing over the sternomastoid muscles and beneath the platysma. They empty into the right and left **subclavian veins** in the base of the neck.

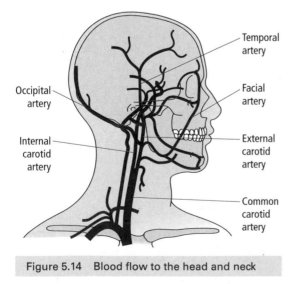

Occipital artery

Internal carotid artery

Temporal artery

Facial artery

External carotid artery

Common carotid artery

Figure 5.14   Blood flow to the head and neck

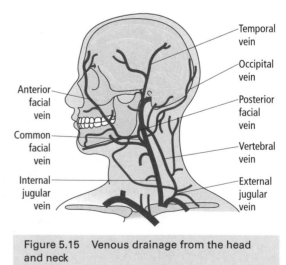

Anterior facial vein

Common facial vein

Internal jugular vein

Temporal vein

Occipital vein

Posterior facial vein

Vertebral vein

External jugular vein

Figure 5.15   Venous drainage from the head and neck

The **internal jugular veins** form the major venous drainage of the head and neck and are deep veins that parallel the **common carotid artery**. They collect deoxygenated blood from the brain and pass downwards through the neck beside the **common carotid arteries** to join the **subclavian veins**.

The **vertebral veins** descend from the transverse openings (or foramina) of the cervical vertebrae and enter the **subclavian veins**. The **vertebral veins** drain deep structures of the neck such as the vertebrae and muscles.

Blood vessels of the arm and hand

The blood supply to the arm begins with the **subclavian artery** (a branch of the aorta).

The subclavian artery becomes the axillary artery and then the **brachial artery**, which runs down the inner aspect of the upper arm to about 1 cm below the elbow where it divides into the **radial** and **ulnar arteries**.

The **radial artery** runs down the forearm and continues over the carpals to pass between the first and second metacarpals into the palm. The **ulnar artery** runs down the forearm next to the ulnar bone, across the carpals into the palm of the hand. Together, the **radial** and **ulnar arteries** form two arches in the hand called the **deep** and **superficial arches**. From these arteries branch others to supply blood to the structures of the upper arm, forearm, hand and fingers.

The venous return of blood from the hand begins with the **palmar arch** and **plexus**, which is a network of capillaries in the palm. The veins that carry deoxygenated blood up

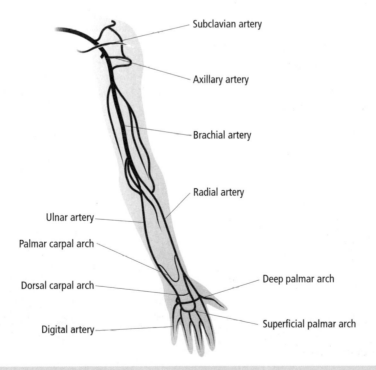

Subclavian artery

Axillary artery

Brachial artery

Radial artery

Ulnar artery

Palmar carpal arch

Dorsal carpal arch

Digital artery

Deep palmar arch

Superficial palmar arch

Figure 5.16   Arteries of the arm and hand

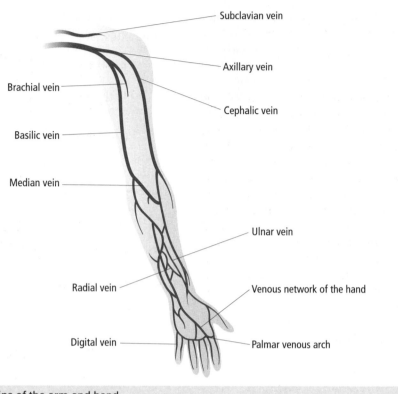

Figure 5.17   Veins of the arm and hand

the forearm are the **radial vein**, **ulnar vein** and **median vein**.

The **radial vein** runs parallel to the radius bone of the forearm; the ulnar vein runs parallel to the ulna bone of the forearm; and the **median vein** runs up the middle of the forearm. Just above the elbow, the **radial** and **ulnar veins** join to become the **brachial vein**, and the **median vein** joins the **basilic vein** which originates just below the elbow along with the **cephalic vein**.

As the veins continue over the elbow they link to form a network that eventually divides, with the **basilic vein** joining the **brachial vein** which then becomes the **axillary vein**. The **cephalic vein** travels up the arm separately and becomes the **subclavian vein** in the upper chest.

Blood vessels of the thoracic and abdominal walls

The thoracic wall is supplied by branches of the **subclavian artery** and the **thoracic aorta**. The abdominal wall is supplied by branches of the **abdominal aorta**. The thoracic and abdominal walls are drained by branches of the **brachiocephalic veins**. Blood from the abdominal organs enters the **hepatic portal system**, and from the liver the blood is carried by the **hepatic veins** to the **inferior vena cava**.

Blood vessels of the leg and foot

The aorta travels down the length of the trunk to the lower abdomen where it divides into two arteries which supply either leg. The **femoral artery** is the artery in the thigh, named after the thigh bone. At the knee the **femoral artery** becomes the **popliteal artery**, which divides into two below the knee. One of these arteries runs down the front of the lower leg and is called the **anterior tibial artery**, while the other runs down the back and is called the **posterior tibial artery**. This artery divides at the inside of the ankle, becoming the **medial plantar artery** on the inside of the foot and the **plantar arch** on the sole of the foot. The **anterior tibial artery** becomes the **dorsal metatarsal artery** on top of the foot.

There is a network of veins in the foot that becomes the **dorsal venous arch** on top of the foot. This travels through the inside of the foot to the ankle where it becomes the **small saphenous vein**. It continues up the back of the whole leg to the thigh where it is known as the **great saphenous vein**.

Two small veins called the **anterior tibial veins** travel up the front of the lower leg, while two veins, the **posterior tibial veins**, run up the back. These four veins converge just below the knee to become the **popliteal vein** at the back of the knee and then eventually the **femoral vein** in the thigh. The **great saphenous vein** and the **femoral vein** join at the groin and return to the heart via the **inferior vena cava**.

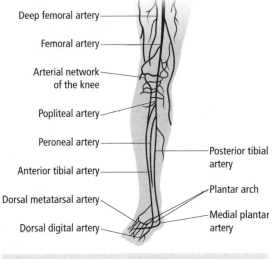

Figure 5.18   Arteries of the leg and foot

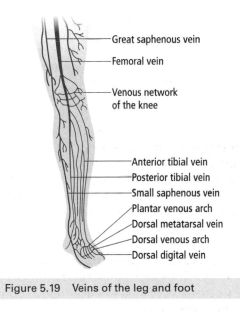

Figure 5.19   Veins of the leg and foot

Blood shunting

Along certain circulatory pathways, such as in the intestines, there are strategic points where small arteries have direct connection with veins. When these connections are open, they act as shunts which allow blood in the artery to have direct access to a vein.

These interconnections allow for sudden and major diversions of blood volume according to the physical needs of the body. In relation to circulation, this means that treatment should not be given after a heavy meal due to the increased circulation to the intestines, resulting in a diminished supply to other areas of the body.

Blood pressure

Blood pressure is the amount of pressure exerted by blood on an arterial wall due to the contraction of the left ventricle. The pressure in the arteries varies during each heartbeat. The maximum pressure of the heartbeat is known as the systolic pressure and represents the pressure exerted on the arterial wall during active ventricular contraction. Systolic pressure can be measured when the heart muscle contracts and pushes blood out into the body through the arteries.

The minimum pressure, or diastolic pressure, represents the static pressure against the arterial wall during rest or pause between contractions. Therefore, the minimum pressure is when the heart muscle relaxes and blood flows into the heart from the veins. Blood pressure may be measured with the use of a sphygmomanometer.

Key note

Blood pressure is regulated by sympathetic nerves in the arterioles. An increase in stimulation of the sympathetic nervous system, as in exercise, can therefore result in a temporary increase in blood pressure.

Factors affecting blood pressure

As blood pressure is the result of the pumping of the heart in the arteries, anything that makes the heart beat faster will raise the blood pressure. Factors affecting the blood pressure include:

- excitement
- pain
- anger
- exercise
- stress
- smoking and drugs.
- fright

A normal blood pressure reading is between 100 and 140 mmHg systolic and between 60 and 90 mmHg diastolic. Blood pressure is measured in millimeters of mercury and is expressed as 120/80 mmHg.

The pulse

The pulse is a pressure wave that can be felt in the arteries which corresponds to the beating of the heart. The pumping action of the left ventricle of the heart is so strong that it can be felt as a pulse in arteries a considerable distance from the heart. The pulse can be felt at any point where an artery lies near the surface. The radial pulse can be found by placing two or three fingers over the radial artery below the thumb. Other sites where the pulse may be felt include the carotid artery at the side of the neck and over the brachial artery at the elbow.

The average pulse in an adult is between 60 and 80 beats per minute. Factors affecting the pulse rate include:

■ exercise

■ heat

■ strong emotions such as grief, fear, anger or excitement.

Disorders of the circulatory system

Anaemia

This is a condition where the haemoglobin level in the blood is below normal. The main symptoms are excessive tiredness, breathlessness on exertion, pallor and poor resistance to infection. There are many causes of anaemia. It may be due to a loss of blood resulting from an accident or operation, chronic bleeding, iron deficiency or a blood disease such as leukaemia.

Aneurysm

Aneurysm is an abnormal, balloonlike swelling in the wall of an artery. This may be due to degenerative disease (congenital defects, arteriosclerosis) or any condition which causes weakening of the arterial wall such as trauma, infection, hypertension.

Angina

This is a pain in the left side of the chest and usually radiating to the left arm, caused by insufficient blood to the heart muscle, usually on exertion or excitement. The pain is often described as constricting or suffocating and can last for a few seconds or moments. Patient may become pale and sweaty. This condition indicates ischaemic heart disease.

Arteriosclerosis

Arteriosclerosis is a circulatory system condition characterised by a thickening, narrowing, hardening and loss of elasticity of the walls of the arteries.

Congenital heart disease

This is a defect in the formation of the heart which usually decreases its efficiency. Defects may be in the following forms:

■ **Ventricular septal defects** – non-closure of the opening between the right and left ventricles

■ **Atrial septal defect** – non-closure of the opening between the right and left atria

■ **Coaraction of the aorta** – narrowing of the aorta

■ **Pulmonary stenosis** – narrowing of the pulmonary artery

■ **Patent ductus arteriosus** – non-closure of the communication between the pulmonary artery and the aorta that exists in the foetus until delivery

■ a combination of defects.

The symptoms may vary according to the severity of the defect.

Haemophilia

This is a hereditary disorder in which the blood clots very slowly due to deficiency of either of two coagulation factors – Factor VIII (the antihaemophiliac factor) or Factor IX (the Christmas factor). They are both coagulation factors normally present in blood. Deficiency of either of these factors, which are inherited by males from their mothers, results in the inability of the blood to clot (haemophilia).

The patient may experience prolonged bleeding following an injury or wound and in severe cases there is spontaneous bleeding into the muscles and joints. Haemophilia is controlled by a sex-linked gene which means it is almost exclusively restricted to males. Women can carry the disease and pass it on to their sons without being affected themselves.

Haemorrhoids (piles)

This is an enlargement of the normal spongy, blood-filled cushions in the walls of the anus. They usually form as a result of prolonged constipation.

Heart attack (myocardial infarction)

This is damage to the heart muscles which results from blockage of the coronary arteries. It can cause serious complications, including heart failure.

Hepatitis

Hepatitis is an inflammation of the liver caused by viruses, toxic substances or immunological abnormalities.

- **Hepatitis A** – this is highly contagious and is transmitted by the faecal–oral route. It is transmitted by ingestion of contaminated food, water or milk. The incubation period is from 15 to 45 days.

- **Hepatitis B** – this is also known as serum hepatitis and is more serious than hepatitis A. It lasts longer and can lead to cirrhosis, cancer of the liver and a carrier state. It has a long incubation period of one and a half to two months. The symptoms may last from weeks to months. The virus is usually transmitted through infected blood, serum or plasma; however, it can spread by oral or sexual contact as it is present in most body secretions.

- **Hepatitis C** – this can cause acute or chronic hepatitis and can also lead to a carrier state and liver cancer. It is transmitted through blood transfusions or exposure to blood products.

High blood pressure

High blood pressure is when the resting blood pressure is above normal. The World Health Organization defines high blood pressure as consistently exceeding 160 mmHg systolic and 95 mmHg diastolic. High blood pressure is a common complaint and, if serious, may result in a stroke or a heart attack, due to the fact that the heart is made to work harder to force blood through the system. Causes of high blood pressure include:

- smoking

- obesity

- lack of regular exercise

- eating too much salt

- excessive alcohol consumption

- too much stress.

High blood pressure can be controlled by:

- antihypertensive drugs which help to regulate and lower blood pressure

- decreasing salt and fat intake to prevent hardening of the arteries

- keeping weight down

- giving up smoking and cutting down on alcohol consumption

- relaxation and leading a less stressful life.

High cholesterol

Cholesterol is a fatlike material present in the blood and most tissues. A high level of cholesterol in the blood (due to a diet rich in animal fats and refined sugars) is often associated with the degeneration of the walls of the arteries and a predisposition to thrombosis.

Low blood pressure

Low blood pressure is when the blood pressure is below normal and is defined by the World Health Organization as a systolic blood pressure of 99 mmHg or less and a diastolic of less than 59 mmHg. Low blood pressure may be normal for some people in good health, during rest and after fatigue. The danger with low blood pressure is an insufficient supply of blood reaching the vital centres of the brain. Treatment may be by medication, if necessary.

Key note

High and low blood pressure are normally contraindicated to treatments, but with GP referral and an adaptation of routine, treatment may be possible. Correct positioning of a therapy couch is essential to maximise comfort of the client with blood pressure problems and care needs to be taken to ensure that they are not lying down too long or get up too fast.

Leukaemia

This term refers to any of a group of malignant diseases in which the bone marrow and other blood-forming organs produce an increased number of certain types of white blood cells. Overproduction of these white cells, which are immature or of abnormal form, suppresses the production of normal white cells, red cells and platelets, which leads to increased susceptibility to infection. Other manifestations or signs include enlargement of the spleen, liver and the lymph nodes, spontaneous bruising and anaemia.

Pacemaker

This is an artificial electrical device implanted under the skin that stimulates and controls the heart rate by sending electrical stimuli to the heart. It is usually installed for heart block and mostly placed in one side of the upper chest.

Phlebitis

This condition is an inflammation of the wall of a vein, which is most commonly seen in the legs, as a complication of varicose veins. A segment of the vein becomes tender and painful and the surrounding skin may feel hot and appear red. Thrombosis may develop as a result of phlebitis (thrombophlebitis) with subsequent deep-vein thrombosis (DVT). This can cause clots in the lungs or other organs with serious consequences.

Pulmonary embolism

This is a blood clot carried into the lungs where it blocks the flow of blood to the pulmonary tissue. This is a very serious condition and can be life-threatening. People who suffer from this condition may require hospitalisation and measures to thin the blood, such as using warfarin. This condition presents with chest pain, cough and shortness of breath.

Raynaud's syndrome

This is a disorder of the peripheral arterioles, characterised by spasm in the smooth muscle of the fingers and toes. It is generally brought on by cold or emotional upset. The effect is a pallor or discolouration of the skin due to the presence of poorly oxygenated haemoglobin. Extremities affected can become painful and uncomfortable and this is usually followed by redness and stiffness of the toes and fingers.

Stress

Stress can be defined as any factor which affects physical or emotional health. When the body is under stress the heart beats faster, increasing the circulation of blood. Excessive or prolonged stress can lead to high blood pressure, coronary thrombosis and heart attack.

Stroke

This is a blocking of blood flow to the brain by an embolus in a cerebral blood vessel.

A stroke can result in a sudden attack of weakness affecting one side of the body, due to the interruption to the flow of blood to the brain. A stroke can vary in severity from a passing weakness or tingling in a limb to a profound paralysis and a coma if severe. Sometimes the term is used to describe cerebral haemorrhage when an artery or congenital cyst of blood vessels in the brain bursts, resulting in damage to the brain and causing similar signs to thrombus (blood clot) of cerebral vessels. Haemorrhage is usually associated with severe headaches and can cause neck stiffness.

Thrombosis

This is a condition in which the blood changes from a liquid to a solid state and produces a blood clot. Thrombosis in the wall of an artery obstructs the blood flow to the tissue it supplies. In the brain this is one of the causes of a stroke, and in the heart it results in a heart attack (coronary thrombosis). Thrombosis may also occur in a vein (deep-vein thrombosis). The thrombus (blood clot) may be detached from its site of formation and be carried in the blood to lodge in another part. (See pulmonary embolism.)

Varicose veins

Veins are known as varicose when the valves within them lose their strength. As a result of this, blood flow may become reversed or static. Valves are concerned with preventing the backflow of blood. When their function is impaired they are unable to prevent the blood from flowing downwards, causing the walls of the affected veins to swell and bulge out and become visible through the skin. Varicose veins may be due to several factors:

- hereditary tendencies
- ageing
- obesity, such as excess weight, putting pressure on the walls of the veins
- pregnancy
- sitting or standing motionless for long periods of time, causing pressure to build up in the vein.

Key note

Varicose veins can be extremely painful and great care needs to be taken with a client with varicose veins. Treatment is, therefore, contraindicated in the area affecting the veins.

Interrelationships with other systems

The cardiovascular system links to the following body systems:

Skin

The circulatory system transports blood rich in nutrients and oxygen to the skin, hair and nails.

Skeletal

Red bone marrow is responsible for the development of blood cells.

Muscular

The heart is a muscular organ and contracts rhythmically and continuously to pump blood around the body.

Lymphatic

The lymphatic system assists the circulatory system in transporting additional waste products away from the tissues in order to maintain blood volume, pressure and prevent oedema.

Respiratory

The respiratory system oxygenates and deoxygenates blood in the lungs.

Nervous

Blood pressure is regulated by sympathetic nerves in the arterioles.

Endocrine

Hormones are carried by blood to their target organs.

Digestive

Nutrients broken down by digestive processes are transported by blood to the liver to be assimilated by the body.

Key words associated with the cardiovascular system

blood	myocardium	cardiac cycle
erythrocyte	endocardium	pulmonary circulation
leucocyte	septum	systemic circulation
thrombocyte	atrium	blood pressure
clotting	ventricle	systolic
artery	superior vena cava	diastolic
vein	inferior vena cava	high blood pressure
capillary	pulmonary artery	low blood pressure
heart	pulmonary veins	pulse
pericardium	aorta	

Summary of the cardiovascular system

- **Blood** is a type of liquid connective tissue.

- Blood transports substances between the body cells and the external environment to help maintain a stable cellular environment.

- The percentage composition of blood is 55 per cent fluid or plasma and 45 per cent blood cells.

- There are three main types of blood cells – **erythrocytes**, **leucocytes** and **thrombocytes**.

- The function of an **erythrocyte** is transporting oxygen to the cells and carrying carbon dioxide away.

- **Leucocytes** are designed to protect the body against infection.

- **Thrombocytes** are involved in the clotting process.

- There are four main functions of blood – transport, defence, regulation of heat and clotting.

- Blood is carried around the body in vessels known as **arteries**, **veins** and **capillaries**.

- **Arteries** carry oxygenated blood away from the heart. They have thick, muscular walls in order to withstand the high pressure of blood.

- **Veins** carry deoxygenated blood towards the heart. They have thinner, muscular and elastic walls and blood is carried under lower pressure.

- **Capillaries** are the smallest vessels in the circulatory system.

- **Capillaries** unite **arterioles** and **venules**. Their walls are sufficiently thin to allow dissolved substances in and out of them.

- The **heart** lies in the **thorax** above the diaphragm and between the lungs.

- The heart is composed of three layers of tissue – an outer **pericardium**, middle **myocardium** and inner **endocardium**.

- The heart is divided into a right and left side by a partition called a **septum**. Each side is divided into a thin-walled top chamber called an **atrium** and a thick-walled bottom layer called a **ventricle**.

- The **atria** (top chambers) take in blood from the large veins and pump it to the bottom chambers.

- The **ventricles** (bottom chambers) pump blood to the body's organs and tissues.

- Blood flows through the heart in three stages.

- In stage 1 deoxygenated blood empties into the **right atrium** from the two main veins (**superior** and **inferior vena cava**). It then flows through the **right ventricle**.

- In stage 2 the **right ventricle** contracts and pushes blood into the **pulmonary artery** and up to the lungs to become oxygenated.

- In stage 3 (occurring at the same time as stage 1) oxygenated blood leaves the **left atrium**, passes through the **left ventricle** and then into the **aorta** and around the body.

- The **cardiac cycle** is the sequence of events between one heartbeat and the next.

- The duration of a **cardiac cycle** is less than a second.

- During a cardiac cycle the **atria** contract simultaneously and force blood into the relaxed ventricles.

- The **ventricles** contract strongly and push blood out through the aorta and the **pulmonary artery**.

- As the **ventricles** contract the **atria** relax and fill up with blood.

- Blood is transported as part of a double circuit.

- The **pulmonary circulation** is the circulatory system between the heart and the lungs. It consists of the circulation of deoxygenated blood from the **right ventricle** of the heart to the **lungs** via the **pulmonary arteries** to become oxygenated. Oxygenated blood is then returned to the **left atrium** by the **pulmonary veins**.

- The **systemic circulation** is the largest circulatory system and carries oxygenated blood from the left ventricle of the heart to the aorta and around the body.

- **Blood pressure** is defined as the amount of pressure exerted by blood on an arterial wall due to the contraction of the **left ventricle**.

- The maximum pressure is called the **systolic** pressure and represents the pressure exerted on the arterial walls during ventricular contraction. The lowest pressure is called the **diastolic** pressure and is when the heart muscle relaxes (ventricular relaxation) and blood flows into the heart from the veins.

- A normal blood pressure reading is between 100 and 140 mmHg **systolic** and between 60 and 90 mmHg **diastolic**.

- **High blood pressure** is when the resting blood pressure is above normal and when consistently exceeding 160 mmHg systolic and 95 mmHg diastolic.

- **Low blood pressure** is defined as a systolic pressure of 99 mmHg or less and diastolic of 59 mmHg.

- The **pulse** is a pressure wave that can be felt in the arteries, such as the carotid or brachial, and corresponds to the beating of the heart and the contraction of the left ventricle.

- An average **pulse** is between 60 and 80 beats per minute.

Multiple-choice questions

1. 55 per cent of the composition of blood is made up of:

a hormones
b haemoglobin
c fluid or plasma
d erythrocytes, leucocytes and thrombocytes

2. The blood cell designed to protect the body against infection is:

a thrombocyte
b leucocyte
c erythrocyte
d platelet

3. Which of the following is not a function of blood?

a transport of oxygen, carbon dioxide, nutrients and hormones
b protection and defence
c synthesis of vitamins A, D and E
d clotting

4. The function of an artery is to:

a carry oxygenated blood
b carry blood under high pressure
c carry blood away from the heart
d all of the above

5. Which of the following statements is false?

a veins carry deoxygenated blood
b veins are generally superficial
c veins do not have valves
d veins carry blood towards the heart

6. The function of a capillary is to:

a carry only deoxygenated blood
b carry only oxygenated blood
c prevent backflow of blood
d supply cells and tissues with nutrients

7. The blood vessel that carries deoxygenated blood from the heart to the lungs is the:

a pulmonary vein
b aorta
c pulmonary artery
d inferior vena cava

8. Oxygenated blood is carried around the body through the various branches of the:

a left ventricle
b left pulmonary veins
c aorta
d superior vena cava

9. Which of the following best describes the flow of blood through the heart?

a blood flows from the capillaries to the veins, to the arteries and then to the aorta
b blood flows into the right atrium, then into the right ventricle, then into the pulmonary artery. The blood returns from the lungs and enters the left atrium, then flows into the left ventricle and into the aorta to all parts of the body
c blood flows into the sinoatrial nodes, then to the right ventricle, aorta and coronary arteries
d blood flows from the brachial artery, to the left ventricle, the pulmonary arteries and on to the right atrium

10. The sounds created by the beating heart are due to:

a contraction of the ventricles
b closing of the heart's valves
c blood moving from one heart chamber to another
d compression from the respiring lungs

11. An average pulse is between:

a   50 and 70 beats per minute
b   60 and 80 beats per minute
c   90 and 120 beats per minute
d   40 and 50 beats per minute

12. A normal blood pressure reading is between:

a   120 and 160 mmHg systolic and between 60 and 90 mmHg diastolic
b   100 and 180 mmHg systolic and between 70 and 80 mmHg diastolio
c   120 and 140 mmHg systolic and between 60 and 90 mmHg diastolic
d   150 and 190 mmHg systolic and between 80 and 90 mmHg diastolic

13. A possible cause of high blood pressure is:

a   smoking
b   obesity·
c   stress
d   all of the above

14. A circulatory disorder in which there is hardening of the arteries is:

a   aneurysm
b   thrombosis
c   angina
d   arteriosclerosis

15. The main vessel that carries deoxygenated blood from the upper part of the body back to heart is:

a   inferior vena cava
b   brachiocephalic vein
c   internal jugular vein
d   superior vena cava

16. Which of the following is true?

a   varicose veins are usually hereditary
b   emotions such as fear decrease the pulse rate
c   the outer layer of the heart is called the myocardium
d   the valve between the right atrium and right ventricle is the bicuspid valve

the lymphatic system and immunity

Introduction

The lymphatic system is a one-way drainage system for the tissues. It helps to provide a circulatory pathway for tissue fluid to be transported as lymph from the tissue spaces of the body into the venous system, where it becomes part of the blood circulation. Through the filtering action of the lymphatic nodes, along with specific organs such as the spleen, the lymphatic system also helps to provide immunity against disease.

Objectives

By the end of this chapter you will be able to recall and understand the following knowledge:

- the functions of the lymphatic system
- the definition of lymph and how it is formed
- the connection between blood and lymph
- the circulatory pathway of lymph
- the names and positions and drainage of the main lymphatic nodes of the head, neck and body
- the immune response
- the interrelationships between the lymphatic and other body systems
- disorders of the lymphatic system.

The human body is equipped with a variety of defence mechanisms that prevent the entry of foreign agents known as pathogens. This defence is called immunity. When working effectively, the immune system protects the body from most infectious microorganisms. It does this both directly, by cells attacking the microorganisms, and indirectly, by releasing chemicals and protective antibodies.

Functions of the lymphatic system

The lymphatic system is important for the distribution of fluid and nutrients in the body because it drains excess fluid from the tissues and returns to the blood protein molecules which are unable to pass back through the blood capillary walls because of their size.

The lymphatic nodes help to fight infection by filtering lymph and destroying invading microorganisms. Lymphocytes are reproduced in the lymph node and following infection they generate antibodies to protect the body against subsequent infection. Therefore, the lymphatic system plays an important part in the body's immune system.

The lymphatic system also plays an important part in absorbing the products of fat digestion from the villi of the small intestine. While the products of carbohydrate and protein digestion pass directly into the bloodstream, fats pass directly into the intestinal lymphatic vessels, known as lacteals.

What is lymph?

Lymph is a transparent, colourless, watery liquid which is derived from tissue fluid and is contained within lymphatic vessels. It resembles blood plasma in composition, except that it has a lower concentration of plasma proteins. This is because some large protein molecules are unable to filter through the cells forming the capillary walls so they remain in blood plasma. Lymph contains only one type of cell and these are called lymphocytes.

How is lymph formed?

As blood is distributed to the tissues some of the plasma escapes from the capillaries and flows around the tissue cells, delivering nutrients such as oxygen and water to the cells and picking up cellular waste such as urea and carbon dioxide. Once the plasma is outside the capillary and is bathing the tissue cells, it becomes tissue fluid. Some of the tissue fluid passes back into the capillary walls to return to the bloodstream via the veins, and some is collected by lymphatic vessels where it becomes lymph. Lymph is then taken through its circulatory pathway and is ultimately returned to the bloodstream.

The connection between blood and lymph

The lymphatic system is, therefore, often referred to as a secondary circulatory system as it consists of a network of vessels that assist the blood in returning fluid from the tissues back to the heart. In this way, the lymphatic system is a complementary system for the circulatory system. After draining the tissues of excess fluid, the lymphatic system returns this fluid to the cardiovascular system. This helps to maintain blood volume, blood pressure and prevent oedema.

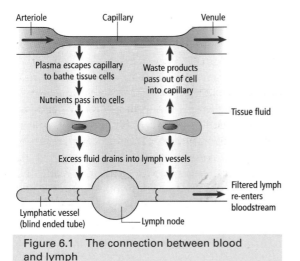

Figure 6.1 The connection between blood and lymph

Structure of the lymphatic system

The lymphatic system contains the following structures:

- lymphatic capillaries
- lymphatic vessels
- lymphatic nodes
- lymphatic collecting ducts.

Lymphatic capillaries

Lymphatic vessels commence as lymphatic capillaries in the tissue spaces of the body as minute blind-end tubes, as lymph is a one-way circulatory pathway. The walls of the lymphatic capillaries are like those of the blood capillaries in that they are the thickness of a single cell layer to make it possible for tissue fluid to enter them. However, they are permeable to substances of larger molecular size than those of the blood capillaries.

The lymphatic capillaries mirror the blood capillaries and form a network in the tissues, draining away excess fluid and waste products from the tissue spaces of the body. Once the tissue fluid enters a lymphatic capillary it becomes lymph and is gathered up into larger lymphatic vessels.

Key note

The term oedema refers to an excess of fluid within the tissue spaces that causes the tissues to become waterlogged.

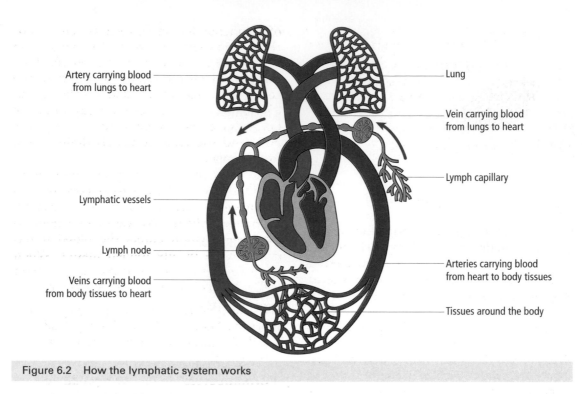

Artery carrying blood from lungs to heart

Lung

Vein carrying blood from lungs to heart

Lymph capillary

Lymphatic vessels

Lymph node

Arteries carrying blood from heart to body tissues

Veins carrying blood from body tissues to heart

Tissues around the body

Figure 6.2   How the lymphatic system works

Lymphatic vessels

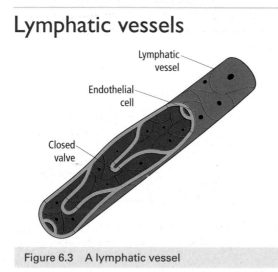

Lymphatic vessel

Endothelial cell

Closed valve

Figure 6.3   A lymphatic vessel

Lymphatic vessels are similar to veins in that they have thin, collapsible walls and their role is to transport lymph through its circulatory pathway. They have a considerable number of valves which help to keep the lymph flowing in the right direction and prevent backflow. Superficial lymphatic vessels tend to follow the course of veins by draining the skin, whereas the deeper lymphatic vessels tend to follow the course of arteries and drain the internal structures of the body. Networks or plexuses of lymphatic channels exist throughout the body. These intertwined channels are found in the following areas:

- **Mammary plexus** – lymphatic vessels around the breasts
- **Palmar plexus** – lymphatic vessels in the palm of the hand
- **Plantar plexus** – lymphatic vessels in the sole of the foot.

The lymphatic vessels carry the lymph towards the heart under steady pressure and about two to four litres of lymph pass into the venous system every day. Once lymph has passed through the lymph vessels it drains into at least one lymphatic node before returning to the blood circulatory system.

Key note

As the lymphatic system lacks a pump, lymphatic vessels have to make use of contracting muscles that assist the movement of lymph. Therefore, lymphatic flow is at its greatest during exercise due to the increased contraction of muscle.

Lymphatic nodes

At intervals along the lymphatic vessels lymphatic nodes occur. A lymphatic node is an oval or bean-shaped structure covered by a capsule of connective tissue. It is made up of lymphatic tissue and is divided into two regions: an outer cortex and an inner medulla.

There are more than 100 lymphatic nodes placed strategically along the course of lymphatic vessels. They vary in size between 1 and 25 mm in length and are massed in groups. Some are superficial and lie just under the skin, whereas others are deeply seated and are found near arteries and veins.

Each lymphatic node receives lymph from several afferent lymphatic vessels and blood from small arterioles and capillaries. Valves

of the afferent lymphatic vessels open towards the node, therefore lymph in these vessels can only move towards the node. Lymph flows slowly through the node, moving from the cortex to the medulla, and leaves through an efferent vessel which opens away from the node. The afferent vessels enter a lymphatic node and the efferent vessels drain lymph from a node.

The function of a lymphatic node is to act as a filter of lymph to remove or trap any microorganisms, cell debris or harmful substances which may cause infection, so that when lymph enters the blood it has been cleared of any foreign matter. When lymph enters a node, it comes into contact with two specialised types of leucocytes:

- **Macrophages** – these are phagocytic in action. They engulf and destroy dead cells, bacteria and foreign material in the lymph.
- **Lymphocytes** – these are reproduced within the lymphatic nodes and can neutralise invading bacteria and produce chemicals and antibodies to help fight disease.

Key note

If an area of the body becomes inflamed or otherwise diseased, the nearby lymph nodes will swell up and become tender, indicating that they are actively fighting the infection.

Once filtered the lymph leaves the node by one or two efferent vessels which open away from the node. Lymphatic nodes occur in chains so that the efferent vessel of one node becomes the afferent vessel of the next node in the pathway of lymph flow. Lymph drains through at least one lymphatic node before it passes into two main collecting ducts before it is returned to the blood.

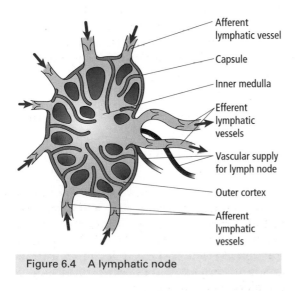

Afferent lymphatic vessel

Capsule

Inner medulla

Efferent lymphatic vessels

Vascular supply for lymph node

Outer cortex

Afferent lymphatic vessels

Figure 6.4   A lymphatic node

Lymphatic ducts

From each chain of lymphatic nodes the efferent lymph vessels combine to form lymphatic trunks which empty into two main ducts – the thoracic and the right lymphatic ducts. These ducts collect lymph from the whole body and return it to the blood via the subclavian veins.

The thoracic duct is the main collecting duct of the lymphatic system. It is the largest lymphatic vessel in the body and extends from the second lumbar vertebra up through the thorax to the root of the neck. The thoracic duct collects lymph from the left side of the head and neck, left arm, lower limbs and abdomen and drains into the left subclavian vein to return it to the bloodstream.

The right lymphatic duct is very short in length. It lies in the root of the neck and collects lymph from the right side of the head and neck and the right arm and drains into the right subclavian vein to be returned to the bloodstream.

Lymphatic drainage

Movement of lymph throughout the lymphatic system is known as lymphatic drainage and it begins in the lymphatic capillaries. The movement of lymph out of the tissue spaces and into the lymphatic capillaries is assisted by:

- The pressure exerted by the skeletal muscles against the vessels during movement.

- Changes in internal pressure during respiration.

- The compression of lymph vessels from the pull of the skin and fascia, during movement.

Key note

Factors such as muscle tension put pressure on the lymphatic vessels and may block them, interfering with efficient drainage. Taking slow, deep breaths can also help to stimulate lymphatic flow.

Lymphatic drainage of the head and neck

The main groups of lymphatic nodes relating to the head and neck are as follows:

Name of lymphatic nodes	Position	Areas from which lymph is drained
Cervical nodes (deep)	Deep within the neck, located along the path of the larger blood vessels (carotid artery and internal jugular vein)	Drain lymph from the larynx, oesophagus, posterior of the scalp and neck, superficial part of chest and arms
Cervical nodes (superficial)	Located at the side of the neck, over the sternomastoid muscle	Drain lymph from the lower part of the ears and the cheek region
Submandibular nodes	Beneath the mandible	Drain chin, lips, nose, cheeks and tongue
Occipital nodes	At the base of the skull	Drain back of scalp and the upper part of the neck
Mastoid nodes (post-auricular)	Behind the ear in the region of the mastoid process	Drain the skin of the ear and the temporal region of the scalp
Parotid nodes	At the angle of the jaw	Drain nose, eyelids and ears

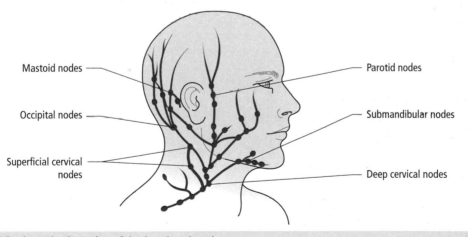

Figure 6.5   Lymphatic nodes of the head and neck

Lymphatic drainage of the body

Lymph nodes are mainly clustered at joints where they assist in pumping lymph through the nodes when the joint moves. The superficial lymph nodes are most numerous in the groin, axillae and neck. Most of the deep lymph nodes are found alongside blood vessels of the pelvic, abdominal and thoracic cavities. The main groups of lymphatic nodes relating to the body are as follows:

Name of lymphatic nodes	Position	Areas from which lymph is drained
Cervical nodes (deep)	Deep within the neck, located along the path of the larger blood vessels	Drain lymph from the larynx, oesophagus, posterior of the scalp and neck, superficial part of chest and arms
Cervical nodes (superficial)	Located at the side of the neck, over the sternomastoid muscle	Drain lymph from the lower part of the ears and the cheek region
Axillary nodes	In the underarm region	Drain the upper limbs, wall of the thorax, breasts, upper wall of the abdomen
Supratrochlear/ cubital nodes	In the elbow region (medial side)	Upper limbs which pass through the axillary nodes
Thoracic nodes	Within the thoracic cavity and along the trachea and bronchi	Organs of the thoracic cavity and from the internal wall of the thorax
Abdominal nodes	Within the abdominal cavity along the branches of the abdominal aorta	Organs within the abdominal cavity
Pelvic nodes	Within the pelvic cavity, along the paths of the iliac blood vessels	Organs within the pelvic cavity
Inguinal	In the groin	Lower limbs, external genitalia and lower abdominal wall
Popliteal	Behind the knee	The lower limbs through deep and superficial nodes

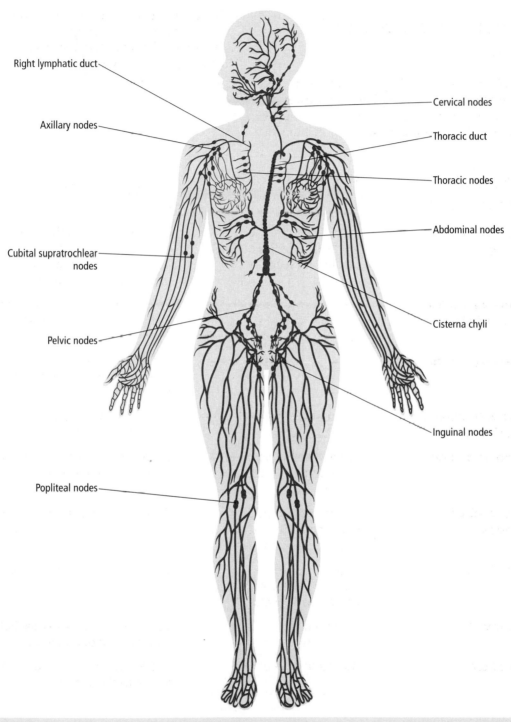

Figure 6.6   Lymphatic nodes of the body

Summary of the circulatory pathway of lymph

- Plasma escapes blood capillary and bathes tissue cells.

- Excess fluid flows through a network of lymphatic capillaries.

- Tissue fluid enters lymph vessels where it becomes lymph.

- Larger lymphatic vessels lead to lymph nodes.

- Lymph passes through at least one lymphatic node where it is filtered.

- Filtered lymph is collected into lymphatic ducts.

- Collected lymph is drained into the venous system via the subclavian veins.

Lymphatic organs

Lymphatic organs, whose functions are closely related to those of the lymph nodes, are the spleen, tonsils and thymus.

Spleen

The spleen is the largest of the lymphatic organs and is located in left-hand side of the abdominal cavity between the diaphragm and the stomach. As the spleen is largely a mass of lymphatic tissue, it contains lymph nodes which produce lymphocytes and macrophages which are phagocytic.

The spleen:

- is a major site for filtering out worn-out red blood cells and destroying microorganisms in the blood.

- is concerned with protection from disease and the manufacture of antibodies. It functions with the lymphatic system by storing lymphocytes and releasing them as part of the immune response.

- serves as a blood reservoir and can release small amounts of blood into the circulation during times of emergency or blood loss.

Tonsils

The tonsils are composed of lymphatic tissue and are located in the oral cavity and the pharynx. There are three different sets of tonsils, all of which provide defence against micro-organisms that enter the mouth and nose.

The palatine tonsils are the set commonly identified as the tonsils and are located at the back of the throat, one on each side. The pharyngeal tonsils are known as the adenoids and lie on the wall of the nasal part of the pharynx. The third set, the lingual tonsils, are found below the tongue.

Thymus

The thymus gland is a triangular-shaped gland composed of lymphatic tissue. It is located in the upper chest above the superior vena cava and below the thyroid, where it lies against the trachea. The function of the thymus is important in the newborn baby in promoting the development and maturation of certain lymphocytes and in programming them to become T-cells (specialised types of lymphocytes of the immune system). The thymus gland begins to atrophy after puberty and becomes only a small remnant of lymphatic tissue in adulthood.

The immune system

The immune system is not a specific structural organ system, but more of a functional system. It draws on the structures and processes of each of the organs, tissues and cells of the body and the chemicals produced in them to eliminate any pathogen, foreign substance or toxic material that can be damaging to the body. Immunity, therefore, can be defined as the ability of the body to resist infection and disease by the activation of specific defence mechanisms.

The human body has a variety of different defence mechanisms. Some are non-specific in that they do not differentiate between one threat and another. Others are specific as the body mounts its defence specifically against a particular kind of threat.

Non-specific immunity

Non-specific immunity is programmed genetically in the human body from birth. The non-specific defences that are present from birth include:

- mechanical barriers
- phagocytosis
- chemicals
- fever
- inflammation.

Mechanical barriers

These are barriers such as the skin and mucous membrane that line the tubes of the respiratory, digestive, urinary and reproductive systems. As long as these barriers remain unbroken, many pathogens are unable to penetrate them.

The respiratory system is lined with mucus-secreting cells to help remove microorganisms from the respiratory tract. The highly acidic environment in the stomach can help to kill pathogens, along with the production of saliva, which has an antimicrobial effect. Urine helps to deter the growth of microorganisms in the genito-urinary tract. The pH of the vagina protects against the multiplication and growth of microbes.

Chemicals

Chemicals are liberated by different cells that play an important role in immunity. There are many different types of chemicals involved in immunity, including interferons, complements and histamine.

Interferons

These are proteins produced by cells infected by viruses. Interferon forms antiviral proteins to help protect uninfected cells and inhibit viral growth. There are three types of human interferon:

- **alpha** (from white blood cells)
- **beta** (from fibroblasts)
- **gamma** (from lymphocytes).

Complements

Complements are proteins found in blood that combine to create substances that phagocytise (ingest) bacteria.

Histamine

This is a chemical released by a variety of tissue cells. This includes mast cells, basophils (a type of white blood cell) and platelets. The release of histamine causes vasodilation to bring more blood to the area of injury or infection. It also increases vascular permeability to allow fluid to enter the damaged area and dilute the toxins released.

Inflammation

Inflammation is a sequence of events involving chemical and cellular activation that destroys pathogens and aids in the repair of tissues. It is a tissue response and symptoms include localised redness, swelling, heat and pain. The major actions that occur during an inflammation response include the following:

- The blood vessels dilate, resulting in an increase in blood volume (hyperaemia) to the affected area.

- Capillary permeability increases, causing tissues to become red, swollen, warm and painful.

- White blood cells invade the area and help to control pathogens by phagocytosis.

- In the case of bacterial infections, pus may form.

- Body fluids collect in the inflamed tissues. These fluids contain fibrinogen and other blood factors that promote clotting.

- Fibroblasts may appear and a connective tissue sac may be formed around the injured tissues.

- Phagocytic cells remove dead cells and other debris from the site of inflammation.

- New cells are formed by cellular reproduction to replace dead or injured ones.

Phagocytosis

Neutrophils and monocytes are the most active phagocytic cells of the blood. Neutrophils are able to engulf and ingest smaller particles, while monocytes can phagocytise larger ones. Monocytes give rise to macrophages (large scavenger cells), which become fixed in various tissues and attached at the inner walls of the blood and lymphatic vessels.

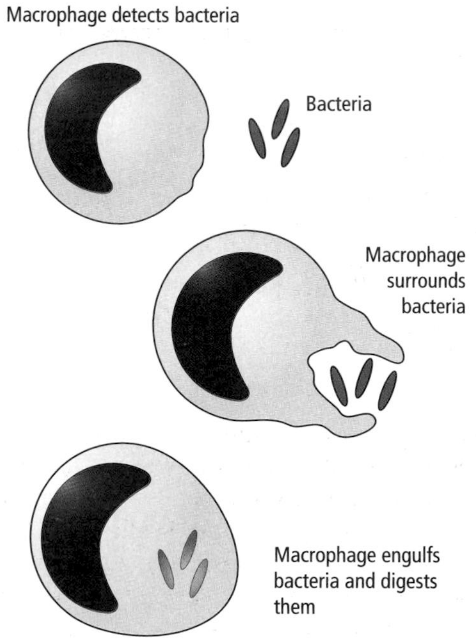

Macrophage detects bacteria

Bacteria

Macrophage surrounds bacteria

Macrophage engulfs bacteria and digests them

Figure 6.7   Phagocytosis

Fever

An individual is said to have a fever if their body temperature is maintained above 37.28°C (99°F). The increase in temperature during a fever tends to inhibit some viruses and bacteria. It also speeds up the body's metabolism and, thereby, increases the activity of defence cells.

Specific immunity

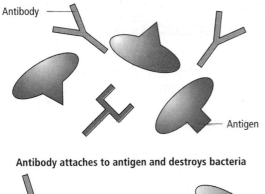

Antibody

Antigen

Antibody attaches to antigen and destroys bacteria

Figure 6.8   The antibody defence system

Immunity involves interaction between two types of molecule – an antigen and an antibody.

An antigen is any substance that the body regards as foreign or potentially dangerous and against which it produces an antibody. An antibody is a specific protein produced to destroy or suppress antigens.

Antibodies circulate in the blood and tissue fluid, killing germs or making them harmless. Antibodies also neutralise poisonous chemicals called toxins which germs produce. Specific immunity involves very specific responses to each identified foreign substance and calls on special memory cells to help if the invader reappears. This is the ability to recognise certain antigens and destroy them. The body must be able to identify which substances are capable of causing a threat before any type of response can be initiated.

How antibodies work

Antibodies work in many different ways. Some neutralise the antigens when they combine with them and prevent them from carrying out their effects. Others may lyse (destroy) the cell on which the antigen is present. When antibodies are bound to antigens on the surface of bacteria, they attract other white blood cells, like macrophages, to engulf them.

The key cells of specific immunity are a specialised group of white blood cells called lymphocytes. They are capable not only of recognising foreign agents, but also of remembering the agents they have encountered, and, therefore, are able to react more rapidly and with greater force if they encounter the agent again.

The immune response

There are two types of immune response produced by different types of lymphocytes:

■ **Humoral immunity** – this involves the B-lymphocytes which produce free antibodies that circulate in the bloodstream.

■ **Cell-mediated immunity** – this is effected by the helper T-cells, suppressor T-cells and natural killer (NK) cells that recognise and respond to certain antigens to protect the body against their effects.

Lymphocytes develop in the following three ways:

■ T-cells begin in the bone marrow and grow in the thymus gland. They are able to recognise antigens and respond by releasing inflammatory and toxic materials. Specialised T-cells also regulate the immune response, either by amplifying the response (T4 cells) or by suppressing the body's response (T8 cells). Some T-cells develop into memory cells and handle secondary response on re-exposure to antigens that have already produced a primary response.

■ B-cells grow and develop in the bone marrow. B-cells contain immunoglobulin, an antibody that responds to specific antigens. Some B-cells modify and become non-antigen-specific, which means that they have a greater ability to respond to bacterial and viral pathogens. Some B-cells become memory cells and are able to deal with re-exposure to antigens.

■ A type of lymphocyte that does not develop the same structural or functional characteristics as the T-cells or B-cells are the natural killer cells (NK) cells. They also develop in the bone marrow and when mature can attack and kill tumour cells and virus-infected cells during their initial developmental stage before the immune system is activated.

Primary and secondary responses

The initial response of the body on first exposure to antigens is known as the primary response. It normally takes about two weeks after exposure to the antigen for antibody levels to peak. This is due to the fact that B-cells have to become converted to plasma cells that secrete antibodies specifically against the antigen.

If the individual is exposed to the antigen the second time, the presence of memory cells stimulates rapid production of antibodies and this is known as the secondary response. The antibody levels are much higher than the primary response and remain elevated for a very long time. Secondary response can occur even if many years have elapsed since the first exposure to the antigen.

Immunisation

The body may be artificially stimulated into producing antibodies and this is known as immunisation. This prepares the body to ward off infection in advance and is carried out by inoculating an individual with a vaccine (a liquid containing antigens powerful enough to stimulate antibody formation without causing harm). Vaccines have been developed against many diseases, including diphtheria, polio, tetanus, whooping cough and measles.

Allergy

Under certain circumstances abnormal responses or allergic reactions may occur when a foreign substance, or antigen, enters the body. An allergic reaction can only occur if the person has already been exposed to the antigen at least once before and has developed some antibody to it.

The type and severity of an allergic reaction depends on the strength and persistence of the antibody screen evoked by previous exposure to the antigen. These antibodies are located on the cells in the skin or mucous membranes of the respiratory and gastro-intestinal tracts. Typical antigens include pollen, dust, feathers, wool, fur, certain foods and drugs.

The reactions may cause symptoms of hay fever, asthma, eczema, urticaria and contact dermatitis. If there is much cellular damage, excessive amounts of histamine may be released, causing circulatory failure (anaphylaxis). Anaphylactic shock is an extreme and generalised form of allergic reaction whose widespread release of histamine causes swelling (oedema), constriction of the bronchioles, heart failure, circulatory collapse and may even result in death.

Disorders of the lymphatic system

Acquired immune deficiency syndrome (AIDS)

This is a condition contracted as a result of the human immunodeficiency virus (HIV) which progressively destroys the immunity of the individual. The HIV virus suppresses the body's immune response, allowing the opportunist infections to take hold and results in AIDS.

AIDS patients become vulnerable to infections that do not affect normal individuals. Infections that produce mild symptoms otherwise may produce severe symptoms. Patients may also be prone to usual cancers. This syndrome is caused by contact with infected blood or body fluids. It is common in drug addicts using infected injection needles and syringes and having unprotected sexual intercourse.

Hodgkin's disease

This is a malignant disease of the lymphatic tissues, usually characterised by painless enlargement of one or more groups of lymph nodes in the neck, armpit, groin, chest or abdomen. The spleen, liver, bone marrow and bones may also be involved. Apart from the enlarging nodes there may also be weight loss, fever, profuse sweating at night and itching.

Lupus erythematosus

This is a chronic inflammatory disease of connective tissue affecting the skin and various internal organs. It is an autoimmune disease and can be diagnosed by the presence of abnormal antibodies in the bloodstream. Typical signs are a red, scaly rash on the face, arthritis and progressive damage to the kidneys. Often the heart, lungs and brain are also affected by progressive attacks of inflammation, followed by the formation of scar tissue. It can also cause psychiatric illness due to direct brain involvement, but the skin is affected in a milder form only.

Oedema

This is an abnormal swelling of body tissues due to an accumulation of tissue fluid. It could be the result of heart failure, liver or kidney disease or due to chronic varicose veins. The resultant swelling of the tissues may be localised, as with an injury or inflammation, or may be more generalised, as in heart or kidney failure. Subcutaneous oedema commonly occurs in the legs and ankles due to the influence of gravity, and is a common problem in women before menstruation and in the last trimester of pregnancy.

Interrelationships with other systems

The lymphatic system links to the following body systems:

Cells and tissues

Lymphatic tissue is a specialised type of tissue found in lymph nodes, spleen, tonsils, the adenoids, walls of the large intestine and glands in the small intestine.

Skin

Lymph vessels are numerous in the dermis of the skin. They form a network allowing the removal of waste from the skin's tissues.

Skeletal

Red bone marrow is responsible for the development of cells found in both blood and lymph.

Muscular

The action of skeletal muscles aids lymphatic drainage.

Circulatory

The lymphatic system aids the circulatory system in that it assists the blood in returning fluid from the tissues back to the heart.

Respiratory

Low pressure in the thorax created by breathing movements aids the movement of lymph.

Digestive

The lymphatic system plays an important part in absorbing the products of fat digestion from the villi of the small intestine.

Key words associated with the lymphatic system

lymph	parotid nodes	tonsils
oedema	axillary nodes	thymus
lacteals	supratrochlear nodes	immunity
lymphatic capillaries	thoracic nodes	specific immunity
tissue (interstitial) fluid	abdominal nodes	non-specific immunity
lymphatic vessels	pelvic nodes	antigen
lymphatic nodes	inguinal nodes	antibody
deep cervical nodes	popliteal nodes	humoral immunity
superficial cervical nodes	thoracic duct	cell-mediated immunity
submandibular nodes	right lymphatic duct	immunisation
occipital nodes	subclavian veins	allergic reaction
mastoid nodes	spleen	

Summary of the lymphatic system

- The lymphatic system is closely associated with the cardiovascular system.

- The lymphatic system assists the blood by draining the tissues of excess fluid and returning the fluid from the tissues back to the heart. This helps to maintain blood volume, blood pressure and prevent **oedema** (waterlogging of the tissues).

- The lymphatic systems also plays an important role in the body's immune system as the lymph nodes fight infection and generate antibodies.

- The lymphatic system also absorbs the products of fat digestion through the intestinal lymph vessels called the **lacteals.**

- **Lymph** is a clear, colourless, watery fluid derived from tissue fluid and contained within lymph vessels.

- Lymph is similar in composition to blood except that it has a lower concentration of plasma proteins.

- The circulatory pathway of lymph begins with **lymphatic capillaries** which lie in the tissue spaces between the cells.

- **Tissue (interstitial) fluid** drains into **lymphatic capillaries** and the excess fluid becomes **lymph.**

- **Lymphatic capillaries** merge to form larger vessels called **lymphatic vessels** which convey lymph in and out of structures called **lymph nodes.**

- The main groups of lymph nodes relating to the head and neck include **deep cervical, superficial cervical, submandibular, occipital, mastoid and parotid nodes**.

- The main group of lymph nodes relating to the body include **superficial cervical, deep cervical, axillary, supratrochlear, thoracic, abdominal, pelvic, inguinal** and **popliteal nodes**.

- Lymph passes through at least one node where it is filtered of cell debris, micro-organisms and harmful substances.

- Once filtered, the lymph is collected into two main ducts – **thoracic duct** (the largest duct), which collects lymph from the left side of the head and neck, left arm, lower limbs and abdomen, and the **right lymphatic duct**, which collects lymph from the right side of the head and neck and the right arm.

- The collected lymph is then drained into the venous system via the right and left **subclavian veins.**

- Other lymphatic organs include the **spleen**, **tonsils** and **thymus gland**.

- **Immunity** is the ability of the body to resist infection and disease by the activation of specific defence mechanisms.

- There are two types of **immunity** – specific and non-specific.

- **Non-specific immunity** is programmed genetically from birth and includes mechanical barriers (skin and mucous membrane), chemicals, inflammation, phagocytosis and fever.

- **Specific immunity** involves interaction between an **antigen** and an **antibody**.

- An **antigen** is any substance that the body regards as foreign or potentially dangerous, and against which it produces an antibody.

- An **antibody** is a specific protein produced to destroy or suppress antigens.

- There are two types of immune response produced by different types of lymphocytes – **humoral immunity** involving B-lymphocytes which produce free antibodies that circulate in the bloodstream, and **cell-mediated immunity** effected by helper T-cells, suppressor T-cells and natural killer (NK) cells that recognise and respond to certain antigens to protect the body against their effects.

- **Immunisation** is when the body is artificially stimulated into producing antibodies.

- An **allergic reaction** may occur when a foreign substance, or antigen, enters the body.

- An allergic reaction can only occur if the person has already been exposed to the antigen at least once before and has developed some antibody to it.

- **Antibodies** are located on the cells in the skin or mucous membranes of the respiratory and gastrointestinal tracts. Typical antigens include pollen, dust, feathers, wool, fur, certain foods and drugs.

Multiple-choice questions

1. Lymph is derived from:

a plasma proteins
b tissue fluid
c blood plasma
d lymphocytes

2. Lymph is similar in composition to blood except it has lower concentration of:

a water
b protein
c waste
d hormones

3. Which of the following is not a function of the lymphatic system?

a production of lymphocytes
b prevention of oedema
c production of heat
d absorption of fat

4. The lymphatic system has a close relationship to which other system?

a nervous
b respiratory
c circulatory
d urinary

5. In order to be cleansed of foreign matter lymph must pass through at least one:

a lymphatic vessel
b lymphatic node
c lymphatic capillary
d lymphatic duct

6. Which of the following statements best describes the structure of a lymph vessel?

a similar to veins
b contain a considerable number of valves
c thin, collapsible walls
d all of the above

7. Lymph flow relies on:

a pressure exerted by skeletal muscles during movement
b compression of lymph vessels from the pull of the skin and fasciae during movement
c changes in internal pressure during respiration
d all of the above

8. Which of the following drains lymph from the lower limbs?

a cervical nodes
b axillary nodes
c popliteal nodes
d supratrochlear nodes

9. Which of the following drains lymph from the back of the scalp and the upper part of the neck?

a occipital
b parotid
c deep cervical
d superficial cervical

10. The axillary nodes are situated in the:

a neck
b groin
c underarm
d elbow

11. The two main lymphatic ducts are the:

a right and left subclavian
b thoracic and left subclavian
c thoracic and right lymphatic
d right and left lymphatic

12. Which duct collects the majority of lymph?

a   thoracic duct
b   left lymphatic duct
c   right lymphatic duct
d   none of the above

13. The largest of the lymphatic organs is the:

a   tonsils
b   spleen
c   liver
d   thymus

14. Collected lymph is drained into the venous system via the:

a   subclavian arteries
b   subclavian veins
c   superior vena cava
d   brachiocephalic veins

15. Which of the following statements is false?

a   an allergic reaction may occur when a foreign substance or antigen enters the body
b   immunity involves interaction between an antigen and an antibody
c   antigens build antibodies on the surface of bacteria
d   the key cells of specific immunity are lymphocytes

16. The viral infection which progressively destroys immunity in an individual is:

a   lupus erythematosus
b   human immunodeficiency virus
c   myalgic encephalomyelitis
d   Hodgkin's disease

the respiratory system

Introduction

The respiratory system consists of the nose, nasopharynx, pharynx, larynx, trachea, bronchi and lungs, which provide the passageway for air in and out of the body. Oxygen is needed by every cell of the body for survival. Respiration is the process by which the living cells of the body receive a constant supply of oxygen and remove carbon dioxide and other gases. Our respiratory system serves us in many ways, exchanging oxygen and carbon dioxide, detecting smell, producing speech and regulating pH.

Objectives

By the end of this chapter you will be able to recall and understand the following knowledge:

- the functions of the respiratory system
- the structure and functions of the main structures of the respiratory system
- the process of the interchange of gases in the lungs
- the mechanism of breathing
- the theory of olfaction
- the importance of correct breathing
- the interrelationships between the respiratory and other body systems
- disorders of the respiratory system.

Functions of the respiratory system

- **Exchange of gases** – oxygen and carbon dioxide exchange is the primary function of the respiratory system in order to sustain life.

- **Olfaction** – specialised nerve endings embedded in the nasal cavity send impulses for the sense of smell to the brain.

- **Speech** – the vocal cords in the larynx aid in producing speech.

- **Homeostasis** – the respiratory system helps to maintain homeostasis by maintaining oxygen levels in the blood and through the elimination of wastes such as carbon dioxide and heat.

The structures of the respiratory system

See Figure 7.1.

The nose

The nose is divided into the right and left cavities. It is lined with tiny hairs called cilia, which begin to filter the incoming air, and mucous membrane, which secretes a sticky fluid called mucus to prevent dust and bacteria from entering the lungs. The nose moistens, warms and filters the air and is an organ which senses smell.

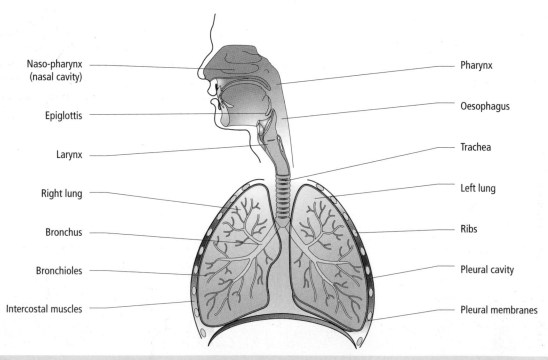

Figure 7.1   Structures of the respiratory system

The nasopharynx

The nasopharynx is the upper part of the nasal cavity behind the nose and is lined with mucous membrane. The eustachian tubes from the middle ears open into the nasopharynx so that air pressure inside the ear can be adjusted to prevent damage to the eardrum. Due to the close proximity of the throat to the eustachian tube, throat infections can easily spread to the ear via the eustachian tubes. At the back of the nasopharynx there is lymphoid tissue such as the adenoids.

The pharynx

The pharynx or throat is a large muscular tube lined with mucous membrane which lies behind the mouth and between the nasal cavity and the larynx. The tonsils are found at the back of the pharynx. The pharynx serves as an air and food passage, but cannot be used for both purposes at the same time, otherwise choking would result. The air is also warmed and moistened further as it passes through the pharynx.

The larynx

The larynx (voice box) is a short passage connecting the pharynx to the trachea. The larynx is a boxlike cavity with rigid walls which contain the vocal cords and stiff pieces of cartilage, such as the Adam's apple, which prevent collapse and obstruction of the airway. The vocal cords are bands of elastic ligaments that are attached to the rigid cartilage of the larynx by skeletal muscle. When air passes over the vocal cords they vibrate and produce sound. The opening into the larynx from the pharynx is called the glottis. During the process of swallowing, the glottis is covered by a flap of tissue called the epiglottis, which prevents food from 'going down the wrong way'. The larynx provides a passageway for air between the pharynx and the trachea.

The trachea

The trachea or windpipe is a tube anterior to the oesophagus and extends from the larynx to the upper chest. It is composed of smooth muscle and up to 20 C-shaped rings of cartilage which serve a dual purpose. The incomplete section of the ring allows the

Key note

The sinuses are air-filled spaces located within the maxillary, frontal, ethmoid and sphenoid bones of the skull. These spaces open into the nasal cavity and are lined with mucous membrane and are continuous with the lining of the nasal cavity. Consequently, mucous secretions can drain from the sinuses into the nasal cavity. If this drainage is blocked by membranes that are inflamed and swollen because of nasal infections or allergic reactions, the accumulating fluids may cause increasing pressure within a sinus and a painful sinus headache. The nasopharynx continues to filter, warm and moisten the incoming air.

oesophagus to expand into the trachea when a food bolus is swallowed and the rings help to keep the trachea permanently open. The trachea passes down into the thorax and connects the larynx with the bronchi which pass into the lungs.

The bronchi

The bronchi are two short tubes, similar in structure to the trachea, which lead to and carry air into each lung. They are lined with mucous membrane and ciliated cells and, like the trachea, contain cartilage to hold them

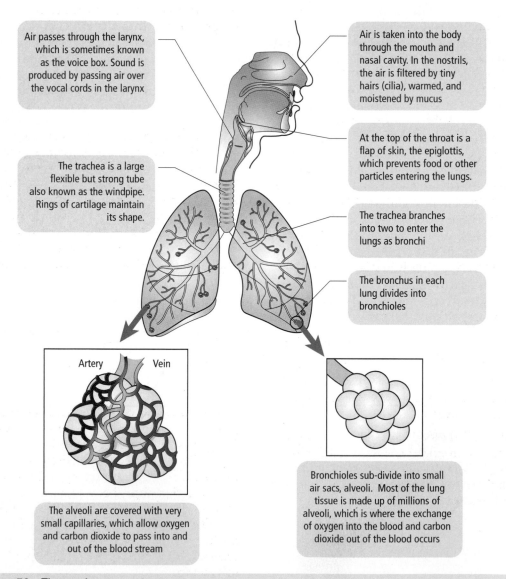

Air passes through the larynx, which is sometimes known as the voice box. Sound is produced by passing air over the vocal cords in the larynx

Air is taken into the body through the mouth and nasal cavity. In the nostrils, the air is filtered by tiny hairs (cilia), warmed, and moistened by mucus

The trachea is a large flexible but strong tube also known as the windpipe. Rings of cartilage maintain its shape.

At the top of the throat is a flap of skin, the epiglottis, which prevents food or other particles entering the lungs.

The trachea branches into two to enter the lungs as bronchi

The bronchus in each lung divides into bronchioles

Artery   Vein

The alveoli are covered with very small capillaries, which allow oxygen and carbon dioxide to pass into and out of the blood stream

Bronchioles sub-divide into small air sacs, alveoli. Most of the lung tissue is made up of millions of alveoli, which is where the exchange of oxygen into the blood and carbon dioxide out of the blood occurs

Figure 7.2   The respiratory system

open. The mucous traps solid particles and cilia move them upwards, preventing dirt from entering the delicate lung tissue. The bronchi subdivide into bronchioles in the lungs. These subdivide yet again and finally end in minute, air-filled sacs called alveoli.

The lungs

The lungs are paired, cone-shaped, spongy organs situated in the thoracic cavity on either side of the heart. The left lung has two lobes and the right lung has three lobes. The right lung is thicker and broader than the left and is also slightly shorter than the left, as the diaphragm is higher on the right side to accommodate the liver which lies below it. Internally, the lungs consist of millions of tiny air sacs called alveoli, which are arranged in lobules and resemble bunches of grapes. The function of the lungs is to facilitate the exchange of the gases oxygen and carbon dioxide. In order to carry this out efficiently, the lungs have several important features:

- a very large surface area (about 1000 square feet) provided by approximately 300 million alveoli

- thin, permeable membrane surrounding the walls of the alveoli

- a thin film of water lining the alveoli which is essential for dissolving oxygen from the alveoli air

- thin-walled blood capillaries forming a network around the alveoli which absorb oxygen from the air breathed into the lungs and release carbon dioxide into the air breathed out of the alveoli.

The structures enclosed within the lungs are bound together by elastic and connective tissue. On the outside the lungs have two layers of a serous membrane called pleura, an outer parietal layer that lines the thoracic cavity and an inner visceral layer that is attached to the surface of the lungs. Between the visceral and parietal pleurae is the pleural cavity, which contains a lubricating fluid secreted by the membranes and reduces friction between the lungs and the chest wall.

The diaphragm

The diaphragm is the chief muscle of respiration and is a dome-shaped, muscular partition that separates the thoracic cavity from the abdominal cavity. During contraction the diaphragm is pulled down, creating a vacuum in the chest cavity which sucks air into the lungs. Relaxation of the diaphragm causes it to rise, allowing the lungs to deflate, and air is pushed out of the lungs as a result.

The interchange of gases in the lungs

Oxygen and carbon dioxide exchange is the primary function of the respiratory system. Oxygen is needed by every cell of the body and delivery is accomplished by way of the bloodstream. The respiratory and circulatory systems, therefore, both participate in this process. The interchange of gases in the lungs involves the absorption of oxygen from the air in exchange for carbon dioxide which is released by the body as a waste product of cell metabolism.

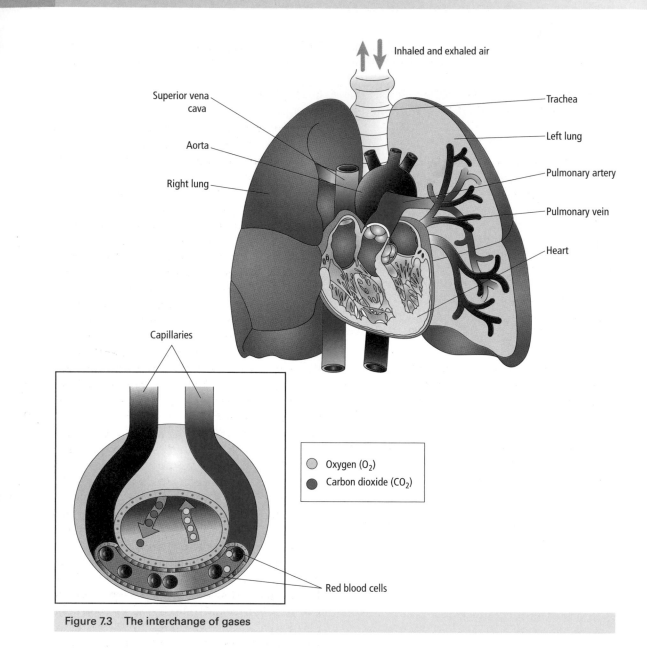

Inhaled and exhaled air

Superior vena cava

Aorta

Right lung

Trachea

Left lung

Pulmonary artery

Pulmonary vein

Heart

Capillaries

○ Oxygen (O$_2$)

● Carbon dioxide (CO$_2$)

Red blood cells

Figure 7.3   The interchange of gases

External respiration

This refers to gas exchange in the lungs between the blood and air in the alveoli that came from the external environment. The respiration process is as follows:

■ During inhalation oxygen is taken in through the nose and mouth. It flows along the trachea and bronchial tubes to the alveoli of the lungs, where it diffuses through the thin film of moisture lining the alveoli.

■ Oxygen diffuses from the air inside the alveoli, across the alveolar walls and into the blood capillaries. The oxygen binds to the haemoglobin inside erythrocytes and is the transported to the cells throughout the body.

■ Carbon dioxide is transported by the blood in the opposite direction from the cells of the body to the capillaries attached to the alveoli.

■ The carbon dioxide then diffuses from the blood, across the alveolar walls, into the air inside the alveoli which will then be exhaled through the nose and mouth.

■ Oxygen and carbon dioxide exchange across the wall of the alveoli at the same time.

Internal/tissue respiration

This is the gas exchange between the blood and the tissues throughout the body. Oxygen diffuses from the blood into the cells and carbon dioxide diffuses from the cells into the bloodstream.

The mechanism of respiration

The mechanism of respiration is the means by which air is drawn in and out of the lungs. It is an active process where the muscles of respiration contract to increase the volume of the thoracic cavity. Air is moved in and out of the lungs by the combined action of the diaphragm and the intercostal muscles.

The major muscle of respiration is the diaphragm. During inspiration the dome-shaped diaphragm contracts and flattens, increasing the volume of the thoracic cavity. It is responsible for 75 per cent of air movement into the lungs. The external intercostal muscles are also involved in respiration, and on contraction they increase the depth of the thoracic cavity by pulling the ribs upwards and outwards. They are responsible for bringing approximately 25 per cent of the volume of air into the lungs. The combined contraction of the diaphragm and the external intercostals increases the thoracic cavity, which then decreases the pressure inside the thorax so that air from outside the body enters the lungs. Other accessory muscles which assist in inspiration include the sternomastoid, serratus anterior, pectoralis minor, pectoralis major and the scalene muscles in the neck.

During normal respiration the process of expiration is passive and is brought about by the relaxation of the diaphragm and the external intercostal muscles, along with the elastic recoil of the lungs. This increases the internal pressure inside the thorax so that air is pushed out of the lungs.

Rib movements in breathing

Inhaling

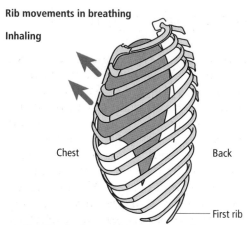

Chest   Back

First rib

Exhaling

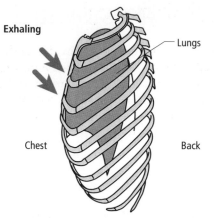

Lungs

Chest   Back

Inhaling. The diaphragm and intercostal muscles contract, pulling the ribs upward. This increases the volume of the chest cavity, drawing air into the lungs.

Exhaling. The contracted muscles relax the ribs fall slightly and decrease the volume of the chest. Air is forced out of the lungs.

How the diaphragm works

Inhaling

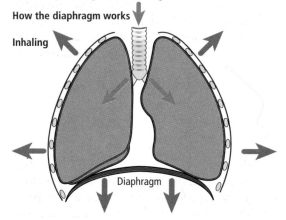

Diaphragm

Exhaling

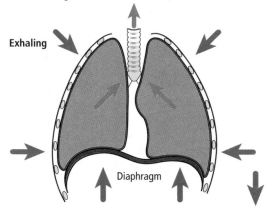

Diaphragm

Inhaling. As the rib cage expands (Arrows, above), the diaphragm contracts and flattens downwards, enlarging the chest cavity.

Exhaling. The diaphragm relaxes and is pressed up by the abdominal organs, returning to its dome shape. The chest narrows, driving air out of the lungs.

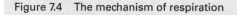

Figure 7.4   The mechanism of respiration

Breathing rate

The normal breathing rate is 12 to 15 breaths per minute, although this may increase during exercise and stress, and decrease during sleep. Breathing takes place rhythmically, with inspiration lasting for about two seconds and expiration for approximately three seconds.

Key note

Breathing is a relatively passive process. However, when more air needs to be exhaled, such as when coughing or playing a wind instrument, the process of expiration becomes active. This is assisted by muscles such as the internal intercostals, which help to depress the ribs. Abdominal muscles, such as the external and internal obliques, rectus abdominis and transversus abdominis, help to compress the abdomen and force the diaphragm upwards, thus assisting expiration and squeezing more air out of the lungs.

Regulation of breathing

Breathing, like the beating of the heart, occurs continuously and rhythmically, without conscious thought. The basic pattern of breathing can be modified by voluntary intervention, but the underlying mechanism is essentially automatic. It continues when we are asleep and unconscious.

Nervous control

Breathing is controlled by a group of neurons in the parts of the brain called the medulla oblongata and the pons, known as the respiratory centre. Nerve cells, called chemo-receptors, found in the aorta and the carotid arteries, send impulses to the respiratory centre in the medulla oblongata of the brain with messages about the levels of oxygen and carbon dioxide in the blood. When the levels of carbon dioxide and oxygen need adjusting, a nerve impulse is sent to the respiratory muscles and, as a result, cellular needs for an adequate supply of oxygen and removal of carbon dioxide are met. The medulla oblongata controls the rate and depth of respiration and the pons moderates the rhythm of the switch from inspiration to expiration.

Olfaction

Olfaction is a special sense which is capable of detecting different smells and evoking emotional responses due to its close link with the endocrine system. The process of olfaction is assisted by the nervous system as smells received by the nose are transmitted by nerve impulses to be perceived by the brain.

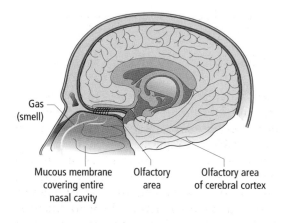

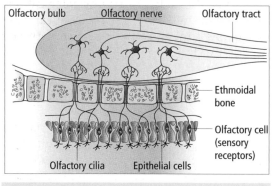

Figure 7.5   Olfaction

The structure of the olfactory system

The special features of the olfactory system are as follows:

- **Nose** – this is the organ of olfaction or smell.

- **Mucous membrane** – this lines the nose, moistens the air passing over it and helps to dissolve the odorous gas passing through the nasal cavity. The mucous membrane has a very rich blood supply, and warmth from the blood flowing

through the tiny capillaries in the nose raises the temperature of the air as it passes through the nose.

- **Cilia** – these are the tiny hairs inside the nose which are covered in mucous. They are highly sensitive and are extensions of nerve fibres connecting with the olfactory cells.

- **Olfactory cells** – these lie embedded in the mucous in the upper part of the nasal cavity. These nerve cells are sensory and are specially adapted for sensing smell. Each olfactory cell has a long nerve fibre called an axon, leading out of the main body of the cell, which picks up information received and passes it on to the brain.

- **Olfactory bulb** – this is the area of the brain situated in the cerebral cortex, which perceives smell.

The theory of olfaction

Olfaction is a special sense in that odour perception is transmitted directly to the brain. The process of olfaction may be summarised as follows:

Reception

The volatile particles of an essential oil evaporate on contact with air. Some volatile molecules pervade the air and some enter the nose. The odiferous particles of the essential oil dissolve in the mucous, which lines the inner nasal cavity, prior to their stimulation at the receptor sites.

Transmission

The captivated aromatic molecules are picked up by the cilia, which protrude from the olfactory receptor cells located at the top of the nasal cavity. The olfactory receptor cells have a long nerve fibre called an axon, and an electrochemical message of the aroma is transmitted along the axons of receptor cells to join the olfactory nerves. The fibres of the olfactory nerves pass through the cribriform plate of the ethmoid bone in the roof of the nose to reach the olfactory bulb, where the odorant signal is chemically converted before being relayed to the brain.

Perception

Once the message reaches the olfactory bulb, the olfactory impulses pass into the olfactory tract and pass directly to the cerebral cortex where the smell is perceived. The temporal lobe of the brain contains the primary olfactory area which is connected directly to the limbic area.

The importance of correct breathing

Exercise increases the rate and depth of breathing due to the muscle cells requiring more oxygen. The breathing rate can more than double during vigorous exercise. Correct breathing is very important as it ensures that all the body's cells receive an adequate amount of oxygen and dispose of enough carbon dioxide to enable them to function efficiently. It is important to note that breathing affects both our physiological and psychological state. Deep breathing exercises can help to increase the vital capacity and function of the lungs.

Key note

In most nerves in the body the transmission of a nerve impulse is achieved through the spinal cord and then the brain. However, in the case of the olfactory cells the nerve fibres connect directly with the olfactory bulb of the brain and, therefore, have a powerful and immediate effect on the emotions. This can be explained by the fact that the area of the brain associated with smell is very closely connected with that part of the brain known as the limbic system, which is concerned with emotions, memory and sex drive. The olfactory bulb also connects closely with the hypothalamus, the nerve centre which governs the endocrine system. Essential oils enter the nose in the form of gases; they evaporate when in contact with air and are volatile in nature. We inhale them in this evaporated form.

Disorders of the respiratory system

Asthma

This condition presents as attacks of shortness of breath and difficulty in breathing due to spasm or swelling of the bronchial tubes. This is caused by hypersensitivity to allergens such as pollens of various plants, grass, flowers, pet hair, dust mites and various proteins in foodstuffs such as shellfish, eggs and milk. Asthma may be exacerbated by exercise, anxiety, stress or smoking. It can run in families and may also be associated with hay fever and eczema.

Bronchitis

This is a chronic or acute inflammation of the bronchial tubes. Chronic bronchitis is common in smokers and may lead to emphysema which is caused by damage to the lung structure. Acute bronchitis can result from a recent cold or flu.

Cancer of the lung

This may be caused by chronic inhalation of cancer-producing air and industrial pollutants such as cigarette smoke and asbestos fibres. Usually there are no symptoms initially and it is often detected only in the advanced stages. Late symptoms include chronic cough, hoarseness, difficulty in breathing, chest pain, blood in sputum, weight loss and weakness.

Emphysema

This is a chronic obstructive pulmonary disease in which the alveoli of the lungs become enlarged and damaged, reducing the surface area for the exchange of oxygen and carbon dioxide. Severe emphysema causes breathlessness which is made worse by infection. It is commonly associated with chronic bronchitis, smoking and advancing age.

Hay fever

This is an allergic reaction involving the mucous passages of the upper respiratory tract and the conjunctiva of the eyes, caused by pollen or other allergens. It causes nose blockages, sneezing and watery eyes.

Pleurisy

This is an inflammation of the pleura of the lung. It presents as an intense, stabbing pain over the chest on breathing deeply. There is difficulty in breathing, respiration is shallow and rapid and fever is present. Pleurisy may develop as a complication of pneumonia, tuberculosis or trauma to the chest.

Pneumonia

Pneumonia is the inflammation of the lung caused by bacteria in which the alveoli become filled with inflammatory cells and the lung becomes solid. Symptoms include fever, malaise and headache, together with a cough and chest pain.

Rhinitis

This condition is the inflammation of the mucous membrane of the nose, causing a blocked, runny and stuffy nose. It may be caused by a virus infection such as a cold, or an allergic reaction.

Sinusitis

This condition involves inflammation of the paranasal sinuses. It is usually caused by a viral or bacterial infection or may be associated with a common cold or allergy. The congestion of the nose results in a blockage in the opening of the sinus into the nasal cavity and a build-up of pressure in the sinus. The condition presents with nasal congestion followed by a mucous discharge from the nose. The pain is located in specific areas depending on the sinuses affected. If the frontal sinuses are affected, a major symptom is a headache over one or both eyes. If the maxillary sinuses are affected, one or both cheeks will hurt and it may feel as if there is a toothache in the upper jaw.

Stress

Stress can be defined as any factor which affects physical or emotional health. Examples of excessive stress on the respiratory system include exacerbation of asthma and the development of frequent colds.

Tuberculosis (TB)

This infectious disease is caused by the bacillus (bacteria) *Mycobacterium tuberculosis.* The main transportation of tuberculosis is via droplet infection, and hence the most common site for the bacilli to spread to is the lungs. The bacilli can also result from drinking unpasteurised milk from infected cows. It is characterised by the formation of nodules in the body tissues. Symptoms include coughing, sneezing, night sweats, fever, weight loss and the spitting of blood. Enlarged lymph nodes can also be an indication of TB. Prevention of the disease is available through the BCG vaccine.

Interrelationships with other systems

The respiratory system links to the following body systems:

Cells and tissues

Squamous and ciliated are examples of types of simple epithelium that line the respiratory system.

Skin

Oxygen absorbed through the respiratory process is carried to the skin via its capillaries to facilitate cell renewal.

Skeletal

The bones of the thorax (sternum, ribs and 12 thoracic vertebrae) provide vital protection for the organs of respiration (heart and lungs).

Muscular

The mechanism of respiration is created by the combined action of the diaphragm and the intercostal muscles.

Circulatory

Blood transports oxygen breathed into the lungs around the body to the cells, and transports carbon dioxide from the cells to the lungs to be exhaled.

Nervous

Breathing is an involuntary response that results from the stimulation of the respiratory centre in the medulla and the pons of the brain.

Endocrine

The hormone adrenalin, produced by the adrenal glands, is released into the bloodstream to change the rate of the breathing when the body is under stress.

Digestive

The mouth and the pharynx link the respiratory and digestive systems.

Key words associated with the respiratory system

nose	bronchioles	olfactory cells
nasopharynx	lungs	olfactory bulb
pharynx	alveoli	diaphragm
larynx	diffusion	intercostal muscles
trachea	cilia	inspiration
bronchi	mucous membrane	expiration

Summary of the respiratory system

- The respiratory organs include the **nose, nasopharynx, pharynx, larynx, trachea, bronchi, bronchioles** and **lungs.**

- The respiratory organs act with the cardiovascular system to supply oxygen and remove carbon dioxide from the blood.

- The **nose** is lined with **cilia** and **mucous membrane** and is adapted for warming, moistening and filtering air and sensing smell.

- Smell is perceived by specialised **olfactory cells** which connect directly with the olfactory bulb in the brain.

- The **pharynx** or throat connects the nasal cavity to the larynx.

- As well as providing an air passage between the nasal cavity and **larynx**, the **pharynx** also serves as a food passage for the digestive system.

- The **larynx** is a short passage that connects the pharynx with the trachea and contains the vocal cords.

- The **trachea** or windpipe is made up mainly of cartilage and passes down into the thorax to connect the larynx with the bronchi which pass into the lungs.

- The **lungs** are situated in the thoracic cavity on either side of the heart.

- Internally the lungs consist of tiny air sacs called **alveoli**, which provide a very large surface area for the exchange of the gases oxygen and carbon dioxide.

- The interchange of gases occurs as a result of simple **diffusion.**

- During **inhalation** oxygen is taken in through the **nose** and **mouth**, along the **trachea** and **bronchi** to the **lungs**, where it diffuses through a thin film of moisture lining the **alveoli.**

- Oxygen then diffuses across the permeable membrane surrounding the **alveoli** to be taken up by the red blood cells, and oxygen-rich blood is carried to the heart and pumped to the cells of the body.

- Carbon dioxide, collected from respiring cells, diffuses from the capillary walls into the **alveoli**, passes through the **bronchi** and **trachea** and is exhaled through the **nose** and **mouth.**

- Air is moved in and out of the lungs by the combined action of the **diaphragm** and the **intercostal muscles.**

- During **inspiration** the combined contraction of the **diaphragm** and the **external intercostals** increases the volume of the **thoracic cavity**, which decreases the pressure inside the thorax so that air enters the lungs.

- The process of **expiration** is passive and is brought about by the relaxation of the **diaphragm** and the **external intercostals** and the elastic recoil of the lungs.

Multiple-choice questions

1. Which of the following is a function of the respiratory system?

a produces speech
b detects smell
c exchanges oxygen and carbon dioxide
d all of the above

2. Which of the following statements is false?

a the pharynx serves as a food and air passage
b the larynx contains the vocal cords
c the pharynx provides a passageway between the larynx and the bronchi
d the bronchi subdivide into bronchioles in the lungs

3. Another name for the throat is the:

a larynx
b pharynx
c epiglottis
d trachea

4. The windpipe is a common name for the:

a pharynx
b epiglottis
c larynx
d trachea

5. The tiny air sacs in the lungs which provide a large surface area for diffusion are:

a surfactants
b alveoli
c pleura
d bronchioles

6. The trachea is made up of mainly:

a spongy tissue
b mucous membrane
c cartilage
d cilia

7. The process of inspiration is brought about by:

a the combined relaxation of the diaphragm and the internal intercostal muscles
b the combined contraction of the diaphragm and the external intercostal muscles
c the combined relaxation of the diaphragm and the external intercostal muscles
d the combined contraction of the diaphragm and the internal intercostal muscles

8. Which of the following statements is true?

a the diaphragm is the chief muscle of respiration
b the diaphragm is responsible for 75 per cent of air movement into the lungs
c breathing is a relatively passive process
d all of the above

9. During gas exchange, oxygen and carbon dioxide diffusion occurs in the:

a red blood cells
b venules
c body tissues
d alveoli

10. When oxygen passes through the alveoli into the bloodstream it binds with haemoglobin to form:

a red blood cells
b carbon dioxide
c nitrogen
d oxyhaemoglobin

11. Involuntary breathing results from stimulation of the respiratory centre in the:

a cerebellum
b thalamus
c medulla and pons
d hypothalamus

12. Which of the following statements is false?

a olfactory cells lie embedded in the mucous in the upper part of the nasal cavity
b smell is perceived by olfactory cells which connect directly with the olfactory bulb in the brain
c the limbic system is concerned with a strong perception of smell
d olfaction can have a powerful and immediate effect on the emotions

the
nervous
system

Introduction

The anatomical structures of the nervous system include the brain, spinal cord and nerves, which together form the main communication system for the body. The nervous system is the body's control centre, or 'head office', and is, therefore, responsible for receiving and interpreting information from inside and outside the body.

The nervous system receives, interprets and integrates all stimuli to effect a response. It is also responsible for all mental processes and emotional responses and works intimately with the endocrine system to help regulate body processes.

Objectives

By the end of this chapter you will be able to recall and understand the following knowledge:

- the functions of the nervous system
- the organisation of the nervous system
- the characteristics of nervous tissue
- the structure and function of different types of neurones
- the transmission of nerve impulses
- an outline of the principal parts of the nervous system
- the interrelationships between the nervous and other body systems
- disorders of the nervous system.

Functions of the nervous system

The nervous system has three main functions:

- It senses changes both within the body (the internal environment) and outside the body (the external environment).

- It analyses the sensory information, stores some aspects and makes decisions as to how to respond. This is called integration.

- It may respond to stimuli by initiating muscular contractions or glandular secretions.

Organisation of the nervous system

The nervous system has two main parts which both possess unique structural and functional characteristics:

- **Central nervous system (CNS)** – this is the main control system that consists of the brain and the spinal cord.

- **Peripheral nervous system (PNS)** – this system can be subdivided into the **somatic nervous system** and the **autonomic nervous system**.

Somatic nervous system

This contains 31 pairs of spinal nerves and 12 pairs of cranial nerves and governs the impulses from the CNS to the skeletal muscles.

Autonomic nervous system

This supplies impulses to smooth muscles, cardiac muscle, skin, special senses, proprioceptors (sensory nerve endings located in muscles and tendons that transmit information to coordinate muscular activity), organs and glands. The autonomic nervous system consists of a sympathetic and parasympathetic division.

Nervous tissue

There are two types of nervous tissue – **neuroglia** and **neurones**.

Neuroglia or glial cells are a special type of connective tissue of the central nervous system that is designed to support, nourish and protect the neurones. Glial cells are smaller and more numerous than neurones. They are unable to transmit impulses and never lose their ability to divide by mitosis.

Key note

Over 50 per cent of the brain is made up of glial cells and most brain tumours are, therefore, made up of glial cells.

The functional unit of the nervous system is a **neurone**, which is a specialised nerve cell, designed to receive stimuli and conduct impulses. The nervous system contains billions of interconnecting neurones, which are the basic impulse-conducting cells of the nervous system. Neurones have two major properties:

- **excitability** – the ability to respond to a stimulus and convert it to a nerve impulse

- **conductibility** – the ability to transmit the impulses to other neurones, muscles and glands.

Neurones also occur in groups called ganglia outside the central nervous system, and as single cells, known as a ganglion, in the walls of organs.

Key note

Nerve cells have the highest metabolic rate in the body and are easily damaged by toxins or lack of oxygen which leads to their destruction.

Parts of a neurone

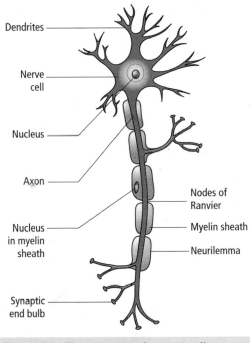

Figure 8.1   The structure of a nerve cell

Although neurones vary in their shape and size they all have three basic parts:

- **Cell body** – this has a central nucleus and is surrounded by cytoplasm, and contains standard organelles such as mitochondria and a Golgi body.
- **Dendrites** – these are highly branched extensions of the nerve cell. These neural extensions receive and transmit stimuli towards the cell body.

- **Axon** – this is long, single nerve fibre extending from the cell body. The function of an axon is to transmit impulses away from the cell body.

Other parts of a neurone's structure include:

Myelin sheath

This is a fatty, insulating sheath that covers the axon. Its function is to insulate the nerve and accelerate the conduction of nerve impulses along the length of the axon. The myelin sheath is produced by Schwann cells (large, flat cells containing a nucleus and cytoplasma) which wrap themselves around the axon in a spiral fashion, layer after layer.

Neurilemma

This is a fine, delicate membrane that surrounds the axon and consists of a layer of one or more Schwann cells enclosing the myelin sheath. The neurilemma plays an important role in the regeneration of PNS nerve fibres.

Nodes of Ranvier

The myelin sheath has gaps at intervals of 2 to 3 mm along the length of the axon which are called the nodes of Ranvier. During neural activity impulses jump from one node to another, resulting in an increased rate of conduction.

Synapse

This is the minute gap across which nerve impulses pass from one neurone to the next at the end of a nerve fibre. Reaching a synapse causes the release of a neurotransmitter which diffuses across the gap and triggers an electrical impulse in the next neurone.

Synaptic end bulb/feet

The ends of the axon terminals have bulb-like structures containing sacs called synaptic vesicles which store the transmitters. These are chemicals which facilitate, arouse or inhibit the transmission of neurones between neurones across synapses.

Key note

A thicker, myelinated nerve fibre will enable nervous signals to be transmitted very quickly, such as pain fibres, whereas hot and cold receptor fibres are non-myelinated and their signals are transmitted more slowly.

There are three types of neurones:

- **Sensory** or afferent neurones receive stimuli from sensory organs and receptors and transmit the impulse to the spinal cord and brain. Sensations transmitted by the sensory neurones include heat, cold, pain, taste, smell, sight and hearing.

- **Motor** or efferent neurones conduct impulses away from the brain and the spinal cord to muscles and glands in order to stimulate them into carrying out their activities.

- **Association** (**mixed**) neurones link sensory and motor neurones, helping to form the complex pathways that enable the brain to interpret incoming sensory messages, decide on what should be done and send out instructions in response, along motor pathways, to keep the body functioning properly.

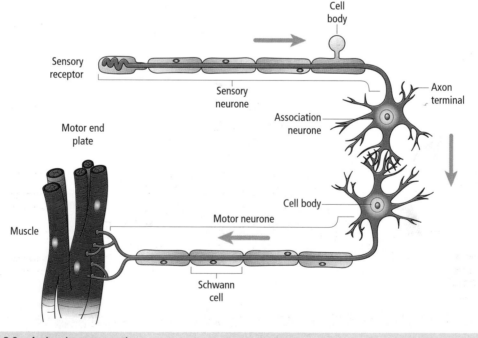

Figure 8.2   A simple nerve pathway

The transmission of nerve impulses

Neurones are responsible for neurotransmission, the conduction of electrochemical impulses throughout the nervous system. Neurone activity is provoked by:

■ **mechanical** stimuli – touch and pressure

■ **thermal** stimuli – heat and cold

■ **chemical** stimuli – from external chemicals or from a chemical released by the body, such as histamine.

Nerve impulses are caused by chemical changes in the cell body. Chemical compounds generate electrical charges called ions. Inside the nerve cell body there are potassium ions which cause a negative charge in the cell, but outside the cell are sodium ions which are positively charged.

Whenever there is change of pressure, temperature or a chemical stimuli, a section of the nerve membrane becomes permeable to sodium and the positively charged ions flow in, leaving the outside of the membrane negative. The combination of the negative potassium ions and the positive sodium ions causes an electrical charge which creates the impulse along the length of the nerve cell.

Nerve impulses are the signals of the nervous system that travel along the neurone from dendrite to axon. The function of a neurone is to transmit impulses from their origin to their destination. The nerve fibres of a neurone are not actually joined together and, therefore, there is no anatomical continuity between one neurone and another.

The junction where nerve impulses are transmitted from one neurone to another is called a synapse. This is the junction between

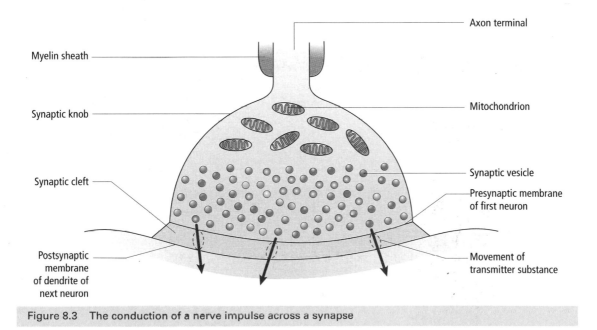

Figure 8.3   The conduction of a nerve impulse across a synapse

two neurones or between a neurone and a muscle or gland where they connect to transmit information.

Impulses are relayed from one neurone to another by a chemical transmitter substance which is released by the neurone to carry impulses across the synapse to stimulate the next neurone. Synapses cause nerve impulses to pass in one direction only and are important in coordinating the actions of neurones.

A special kind of synapse occurs at the junction between a nerve and a muscle and is known as a motor point, which is the point where the nerve supply enters the muscle.

The conduction of a motor impulse in the contraction of skeletal muscle

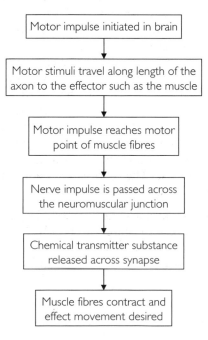

Motor impulse initiated in brain

↓

Motor stimuli travel along length of the axon to the effector such as the muscle

↓

Motor impulse reaches motor point of muscle fibres

↓

Nerve impulse is passed across the neuromuscular junction

↓

Chemical transmitter substance released across synapse

↓

Muscle fibres contract and effect movement desired

Key note

An important neurotransmitter is **acetylcholine**, which is vital to muscle contraction.

The central nervous system

The central nervous system, consisting of the brain and spinal cord, is covered by a special type of connective tissue called the **meninges**. The meninges has three layers:

- **Dura mater** – this is the outer, protective, fibrous, connective, tissue sheath covering the brain and spinal cord.

- **Pia mater** – this is the innermost layer which is attached to the surface of organs and is richly supplied with blood vessels to nourish the underlying tissues.

- **Arachnoid mater** – this provides a space for the blood vessels and circulation of cerebrospinal fluid.

Cerebrospinal fluid

This is a clear fluid derived from the blood and secreted into the inner cavities of the brain. It carries some nutrients to the nerve tissue and carries waste away, but its main function is to protect the central nervous system by acting as a shock absorber for the delicate nervous tissue.

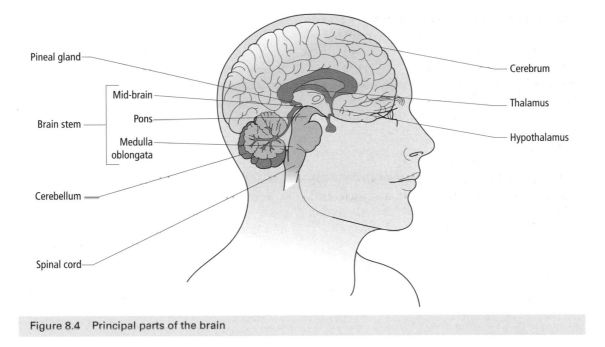

Figure 8.4 Principal parts of the brain

The brain

The brain is an extremely complex mass of nervous tissue lying within the skull. It is the main communication centre of the nervous system and its function is to coordinate the nerve stimuli received and effect the correct responses. The main parts of the brain include the cerebrum, thalamus, cerebellum and brain stem.

Cerebrum

This is the largest portion of the brain and makes up the front and top part of the brain. It is divided into two large cerebral hemispheres. Each cerebral hemisphere is divided into four lobes – **frontal**, **temporal**, **parietal** and **occipital**, named according to the skull bones that lie over them.

A mass of nerve fibres known as the **corpus callosum** bridges the hemispheres, allowing communication between corresponding centres in each hemisphere. The surface of the cerebrum is made up of convolutions called gyri and creases called sulci.

The outer layer of the cerebrum is called the **cerebral cortex** and is the region where the main functions of the cerebrum are carried out. The cortex is concerned with all forms of conscious activity, such as vision, touch, hearing, taste and smell, as well as control of voluntary movements, reasoning, emotion and memory. The cortex of each cerebral hemisphere has a number of functional areas:

- **Sensory areas** – these receive impulses from sensory organs all over the body. There are separate sensory areas for vision, hearing, touch, taste and smell.

- **Motor areas** – these areas have motor connections through motor nerve fibres with voluntary muscles all over the body.

- **Association areas** – in these areas association takes place between information from the sensory areas and remembered information from past experiences. Conscious thought then takes place and decisions are made which often result in conscious motor activity controlled by motor areas.

Key note

The brain requires a continuous supply of glucose and oxygen as it is unable to store glycogen, unlike the liver and muscles.

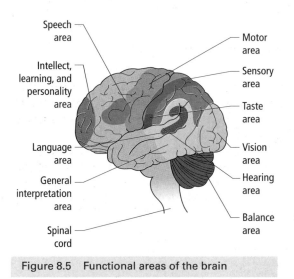

Figure 8.5   Functional areas of the brain

Thalamus

Lying deep in the cerebral hemispheres in each side of the forebrain are one of two egg-shaped masses of grey matter called the thalami. The thalami are relay and interpretation stations for the sensory messages (except olfaction) that enter the brain before they are transmitted to the cortex.

Hypothalamus

This small structure lies beneath the thalamus and governs many important homeostatic functions. It regulates the autonomic nervous and endocrine systems by governing the pituitary gland. It controls hunger, thirst, temperature regulation, anger, aggression, hormones, sexual behaviour, sleep patterns and consciousness.

Pineal gland

This is a pea-sized mass of nerve tissue attached by a stalk in the central part of the brain. It is located deep between the cerebral hemispheres where it is attached to the upper portion of the thalamus. The pineal gland secretes a hormone called melatonin which is synthesised from serotonin. The pineal gland is involved in the regulation of circadian rhythms. These are patterns of repeated activity that are associated with the environmental cycles of day and night such as sleep/wake rhythms. The pineal gland is also thought to influence mood.

Cerebellum

The cerebellum is a cauliflower-shaped structure located at the posterior of the cranium, below the cerebrum. It is the brain's second largest region. Like the cerebrum, it has two hemispheres and has an outer cortex of grey matter and an inner core of white matter. The cerebellum is concerned with muscle tone, the coordination of skeletal muscles and balance.

Brain stem

The brain stem contains three main structures:

- **Midbrain** – this contains the main nerve pathways connecting the cerebrum and the lower nervous system. It also contains certain visual and auditory reflexes that coordinate head and eye movements with things seen and heard.

- **Pons** – this is below the midbrain and relays messages from the cerebral cortex to the spinal cord and helps regulate breathing.

- **Medulla oblongata** – this is often considered the most vital part of the brain. It is an enlarged continuation of the spinal cord and connects the brain with the spinal cord. Control centres within the medulla oblongata include those for the heart, lungs and intestines. The medulla also controls gastric secretions and reflexes such as sweating, sneezing, swallowing and vomiting.

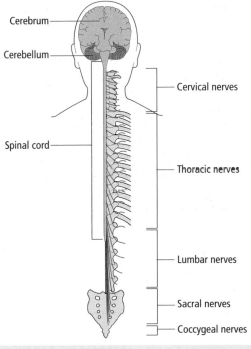

Figure 8.6   The spinal cord

Blood-brain barrier

The blood-brain barrier is a selective semi-permeable wall of blood capillaries with a thick basement membrane. It prevents, or slows down, the passage of some drugs and other chemical compounds and keeps disease-causing organisms such as viruses from travelling into the central nervous system via the bloodstream.

Spinal cord

This is an extension of the brain stem which extends from an opening at the base of the skull down to the second lumbar vertebra.

It forms a two-way information pathway between the brain and the rest of the body via the spinal nerves. It is protected by three layers of tissues called the meninges and by cerebrospinal fluid. Its function is to relay impulses to and from the brain. Sensory tracts conduct impulses to the brain and motor tracts conduct impulses from the brain.

The spinal cord provides the nervous tissue link between the brain and other organs of the body and is the centre for reflex actions which provide a fast response to external or internal stimuli.

Reflex action

A reflex action is a rapid and automatic response to a stimulus without any conscious thought of the brain.

Reflexes are essentially designed to protect the body. A reflex action, sometimes called a reflex arc, is a neural relay cycle for quick motor response to a harmful sensory stimulus. It requires a sensory (afferent) neurone and a motor (efferent) neurone.

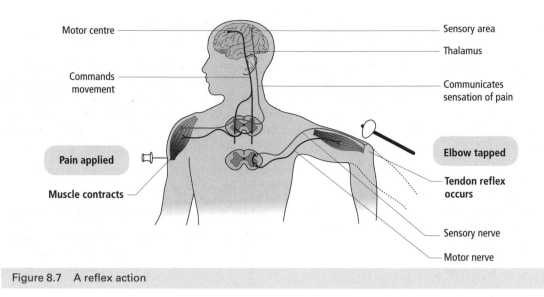

Figure 8.7   A reflex action

Instead of the sensory impulse going all the way to the brain where it can be analysed and the correct response selected, a reflex allows a shorter and quicker response. A typical example of a reflex action is a hand touching a hot object, which involves sensory and motor nerves being coordinated through the spinal cord. The stimulus triggers a sensory impulse which travels along the dorsal root to the spinal cord.

Two synaptic transmissions occur at the same time. One synapse continues the impulse along a sensory neurone to the brain, while the other immediately relays the impulse to an interneurone which transmits it to a motor neurone.

The motor neurone delivers the impulse to a muscle or gland, producing an immediate response, and in this case withdrawing the hand from the hot object.

The peripheral nervous system

The peripheral nervous system contains all the nerves outside the central nervous system. It consists of cablelike nerves that link the central nervous system to the rest of the body. The peripheral nervous system can be subdivided into the somatic nervous system and the autonomic nervous system.

The somatic nervous system contains:

- **31 pairs of spinal nerves** (nerves originating from the spinal cord)
- **12 pairs of cranial nerves** (nerves originating from the brain).

31 pairs of spinal nerves

These nerves pass out of the spinal cord and each has two thin branches which link it with the autonomic nervous system. Spinal nerves receive sensory impulses from the body and transmit motor signals to specific regions of the body, thereby providing two-way communication between the central nervous system and the body.

Each of the spinal nerves are numbered and named according to the level of the spinal column from which they emerge. There are:

- **8 cervical nerves**
- **12 thoracic nerves**
- **5 lumbar nerves**
- **5 sacral nerves**
- **1 coccygeal spinal nerve**.

Each spinal nerve is divided into several branches, forming a network of nerves or plexuses which supply different parts of the body:

- **The cervical plexuses** of the neck supply the skin and muscles of the head, neck and upper region of the shoulders.

- **The brachial plexuses** are at the top of the shoulder and they supply the skin and muscles of the arm, shoulder and upper chest.

- **The lumbar plexuses** are located between the waist and the hip. They supply the front and sides of the abdominal wall and part of the thigh.

- **The sacral plexuses** at the base of the abdomen supply the skin and muscles and organs of the pelvis.

- **The coccygeal plexus** supplies the skin in the area of the coccyx and the muscles of the pelvic floor.

12 pairs of cranial nerves

These nerves connect directly to the brain. Between them they provide a nerve supply to sensory organs, muscles and skin of the head and neck. Some of the nerves are mixed, containing both motor and sensory nerves, while others are either sensory or motor.

- **Olfactory** – this is a sensory nerve of olfaction (smell).

- **Optic** – this is a sensory nerve of vision.

- **Oculomotor** – this is a mixed nerve that innervates both internal and external muscles of the eye and a muscle of the upper eyelid.

- **Trochlear** – this is the smallest of the cranial nerves. It is a motor nerve that innervates the superior oblique muscle of the eyeball which helps you look upwards.

- **Abducens** – this is a mixed nerve that innervates only the lateral rectus muscle of the eye which helps you look to the side.

- **Facial** – this is a mixed nerve that conducts impulses to and from several areas in the face and neck. The sensory branches are associated with the taste receptors on the tongue and the motor fibres transmit impulses to the muscles of facial expression.

- **Vestibulocochlear** – this is a sensory nerve that transmits impulses generated by auditory stimuli and stimuli related to equilibrium, balance and movement.

- **Glossopharyngeal** – this is a mixed nerve that innervates structures in the mouth and throat. It supplies motor fibres to part of the pharynx and to the parotid salivary glands, and sensory fibres to the posterior third of the tongue and the soft palate.

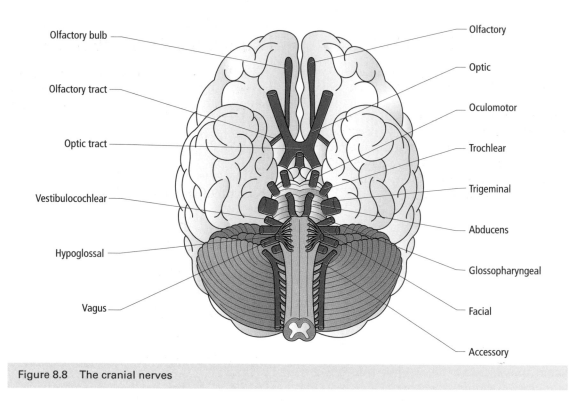

Figure 8.8   The cranial nerves

- **Vagus** – unlike the other cranial nerves, this has branches to numerous organs in the thorax and abdomen as well as the neck. It supplies motor nerve fibres to the muscles of swallowing and to the heart and organs of the chest cavity. Sensory fibres carry impulses from the organs of the abdominal cavity and the sensation of taste from the mouth.

- **Accessory** – this functions primarily as a motor nerve, innervating muscles in the neck and upper back, such as the trapezius and the sternomastoid, as well as muscles of the palate, pharynx and larynx.

- **Hypoglossal** – this is a motor nerve that innervates the muscles of the tongue.

- **Trigeminal** – this is a mixed nerve containing motor and sensory nerves that conducts impulses to and from several areas in the face and neck. It also controls the muscles of mastication (the masseter, temporalis and pterygoids).

The trigeminal nerve has three main branches:

- The **ophthalmic** branch carries sensations from the eye, nasal cavity and skin of the forehead, upper eyelid, eyebrow and part of the nose.

- The **maxillary** branch carries sensations from the lower eyelid, upper lip, gums, teeth, cheek, nose, palate and part of the pharynx.

- The **mandibular** branch carries sensations from the lower gums, teeth, lips, palate and part of the tongue.

The autonomic nervous system

This is the part of the nervous system that controls the automatic body activities of smooth and cardiac muscle and the activities of glands. It is divided into the sympathetic and parasympathetic divisions, which possess complementary responses.

The sympathetic system

The activity of the sympathetic system is to prepare the body for expending energy and dealing with emergency situations. Its effects include:

- increased heart rate
- increased respiration rate
- dilation of skeletal blood vessels

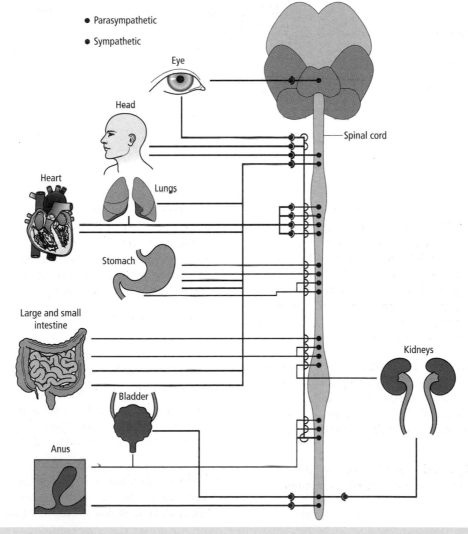

Figure 8.9   The autonomic nervous system

- stimulation of the adrenal and sweat glands

- increased conversion of glycogen to glucose by liver

- pupil dilation

- decreased secretion of saliva

- relaxation of the bladder wall and the closing of the sphincter muscles

- decreased gastrointestinal activity.

Key note

> The sympathetic stimulation of the autonomic nervous system is increased by the release of the hormone adrenaline from the adrenal medulla. This is an example of the nervous and endocrine systems working synergistically.

The parasympathetic system

This balances the action of the sympathetic division by working to conserve energy and create the conditions needed for rest and sleep. It slows down the body processes, except digestion and the functions of the genito-urinary system. In general, the actions of the parasympathetic system oppose those of the sympathetic system and the two systems work together to regulate the internal workings of the body. Effects of the parasympathetic nervous system include:

- resting heart rate

- resting respiratory rate

- constriction of skeletal blood vessels

- increased gastrointestinal activity (increased peristalsis and increased secretion of insulin and digestive juices)

- contraction of the bladder and the opening of the sphincter muscles

- pupil constriction

- stimulated salivation.

The sympathetic and parasympathetic nervous systems are finely balanced to ensure the optimum functioning of organs of the body.

Sense organs

The sense organs include:

- **Nose** (olfaction) – the specialised chemoreceptor olfactory nerve cells in the nose pick up information of an incoming odour and pass it to the olfactory bulb in the brain to be analysed.

- **Tongue** (taste) – chemosensitive receptors are concentrated on the projections on the tongue called papillae. Within the papillae lie the tiny taste buds which are round in structure and form bundles of cell bodies and nerve endings of the seventh, ninth and tenth cranial nerves. The taste hairs are stimulated by food and drink that is placed in the mouth, sending messages in the form of electrical impulses to the taste area in the cerebrum for interpretation.

- **Eyes** (sight) – vision uses photoreceptors that are located in the eye. Light enters the eye through the pupil and strikes the retina. There are two types of photo-receptors located on the retina: rods and cones. These light-sensitive cells convert the incoming light into nerve impulses and send them to the optic nerve to interpret what is being seen.

- **Ears** (hearing) – the ears are associated with the sensory functions of hearing and balance which are detected by mechanoreceptors. Sound waves are picked up in the ear and transmitted to the cerebrum via the eighth cranial nerve for interpretation. The ears send messages via the eighth cranial nerve to the cerebrum and cerebellum to detect changes in the position of the head. The messages are interpreted and the skeletal muscles are instructed to maintain balance and posture.

- **Skin** (touch) – there are numerous sensory nerve endings in the skin that are sensitive to touch, pain and changes in temperature (see Chapter 2).

Disorders of the nervous system

Anxiety

This can be defined as fear of the unknown, but as an illness it can vary from a mild form to panic attacks and severe phobias that can be disabling socially, psychologically and, at times, physically.

It presents with a feeling of dread that something serious is likely to happen and is associated with palpitations, rapid breathing, sweaty hands, tremor (shakiness), dry mouth, general indigestion, feeling of butterflies in the stomach, occasional diarrhoea and generalised aches and pains in the muscles. It can present with similar features of mild-to-moderate depression of the agitated type. The causes of anxiety can be related to personality, with some genetic and behavioural predisposition, a traumatic experience or physical illness such as hyperthyroidism.

Bell's palsy

This is a disorder of the seventh cranial nerve (facial nerve) that results in paralysis on one side of the face. The disorder usually comes on suddenly and is commonly caused by inflammation around the facial nerve as it travels from the brain to the exterior. It may be caused by pressure on the nerve due to tumours, injury to the nerve, infection of the meninges or inner ear or dental surgery. Diabetes, pregnancy and hypertension are other causes.

The condition may present with a drooping of the mouth on the affected side due to flaccid paralysis of the facial muscles, and there may be difficulty in puckering the lips due to paralysis of the orbicularis oris muscle. Other symptoms include:

- Taste may be diminished or lost if the nerve has been affected proximal to the branch which carries taste sensations.
- It may be difficult to close the eye tightly and crease the forehead.
- The buccinator muscle is affected which prevents the person from puffing the cheeks and is the cause of food getting caught between the teeth and cheeks.
- There is excessive tearing from the affected eye.
- Pain may be present near the angle of the jaw and behind the ear.

Between 80 and 90 per cent of individuals recover spontaneously and completely in around one to eight weeks. Corticosteriods may be used to reduce the inflammation of the nerve.

Carpal tunnel syndrome

This syndrome is characterised by pain and numbness in the thumb or hand, resulting from pressure on the median nerve of the wrist. Pain and a pins-and-needles sensation may radiate to the elbow. It is known to cause severe pain at night and can cause muscle wasting of the hand. There is a higher risk of this condition in occupations requiring repetitive strains of the wrist, such as massage therapists and secretaries.

Cerebral palsy

This condition is caused by damage to the central nervous system of the baby during pregnancy, delivery or soon after birth. The damage could be due to bleeding, lack of oxygen or other injuries to the brain. The signs and symptoms of this condition depend on the area of the brain affected. Symptoms include:

- Speech is impaired in most individuals and there may be difficulty swallowing.
- There may or may not be mental retardation.
- Muscles may increase in tone to become spastic, making coordinated movements difficult. The muscles are hyperexcitable and even small movements, touch, stretch of muscle or emotional stress can increase the spasticity.
- The posture is abnormal due to muscle spasticity and the gait is also affected. Some may have abnormal involuntary movements of the limbs that may be exaggerated on voluntarily performing a task. Weakness of muscles may also be associated with the condition, along with seizures.
- There may be problems with hearing and vision.

Depression

This combines symptoms of lowered mood, loss of appetite, poor sleep, lack of concentration and interest, lack of sense of enjoyment, occasional constipation and loss of libido. There are occasions when there is suicidal thinking, death wish or active suicide attempts.

Depression can be the result of chemical imbalance, usually related to serotonin and noradrenaline. The cause of depression could be endogenous, where there is no cause but is thought to be linked to genetic predisposition, the result of physical illness or the loss of a close relative, object, limb or a relationship. A depressed person looks miserable, hunchbacked, downcast and will usually avoid eye contact.

The severity can be variable but may become severe enough to become psychotic, manifested by hallucinations, delusions, paranoia or thought disorders.

Epilepsy

This is a neurological disorder which makes the individual susceptible to recurrent and temporary seizures. Epilepsy is a complex condition and classifications of types of epilepsy are not definitive. Types of epilepsy are:

- **Generalised** – this may take the form of major or tonic-clonic seizures (formerly known as grand mal) in which at the onset the patient falls to the ground, unconscious, with their muscles in a state of spasm (tonic phase). This is then replaced by convulsive movements (clonic phase) when the tongue may be bitten and urinary incontinence may occur. Movements gradually cease and the patient may rouse in a state of confusion, complaining of a headache, or may fall asleep.

- **Partial** – this may be idiopathic or a symptom of structural damage to the brain. In one type of partial idiopathic epilepsy, often affecting children, seizures may take the form of absences (formerly known as petit mal), in which there are brief spells of unconsciousness lasting for a few seconds. The eyes stare blankly and there may be fluttering movements of the lids and momentary twitching of the fingers and mouth. This form of epilepsy seldom appears before the age of three or after adolescence. It often subsides spontaneously in adult life but may be followed by the onset of generalised epilepsy.

- **Focal** – this is partial epilepsy due to brain damage (either local or due to a stroke). The nature of the seizure depends on the location of the damage in the brain. In a Jacksonian motor seizure the convulsive movements may spread from the thumb to the hand, arm and face.

- **Psychomotor** – this type of epilepsy is caused by a dysfunction of the cortex of the temporal lobe of the brain. Symptoms may include hallucinations of smell, taste, sight and hearing. Throughout an attack the patient is in a state of clouded awareness and afterwards may have no recollection of the event.

Headache

This is a pain affecting the head, excluding facial pain. It can result from diseases affecting ear, nose and throat such as sinusitis, as well as eye problems which could be corrected by glasses. Types of headaches include:

- **Simple headache** – this may occur at times of stress, during menstruation, the day after heavy alcohol consumption and as part of cold and flu symptoms. These are transient and would normally settle spontaneously or require simple analgesia.

- **Chronic headaches** – these are daily headaches and tension headaches. The pain can be severe and disabling and can affect the whole head, behind the eyes or may be just a frontal headache. The client can describe the pain as like a band around the head.

- **Cervical spines (cervicalgia)** – this is normally in the back and sides of the head and can present with neck pain.

- **Migraine headache** – this is a specific form of headache, usually unilateral (one side of the head), associated with nausea or vomiting and visual disturbances such as scintillating or zigzag light waves.

- **Intracranial (inside brain) diseases** – these are headaches caused by diseases such as a brain tumour, which can present with nausea and vomiting and may cause other neurological signs and symptoms.

Herpes zoster (shingles)

This is a painful infection along the sensory nerves by the virus that causes chicken pox. Lesions resemble herpes simplex, with erythema and blisters along the lines of the nerves. Areas affected are mostly on the back or upper chest wall. This condition is very painful due to acute inflammation of one or more of the peripheral nerves. Severe pain may persist at the site of shingles for months or even years after the apparent healing of the skin.

Meningitis

This is an inflammation of the meninges due to infection by viruses or bacteria. Meningitis presents with an intense headache, fever, loss of appetite, intolerance to light and sound and rigidity of muscles, especially those in the neck. In severe cases there may be convulsions, vomiting and delirium leading to death. The different types of meningitis are:

- **Meningococcal meningitis** – this involves a characteristic haemorrhagic rash anywhere on the body. The symptoms appear suddenly and the bacteria can cause widespread meningococcal infection, culminating in meningococcal septicaemia. Unless treated rapidly death can occur within a week.

- **Bacterial meningitis** – this is treated with large doses of antibiotics.

- **Viral meningitis** – this does not respond to drugs but normally has a relatively benign prognosis.

Migraine

This is a specific form of headache, usually unilateral (one side of the head), associated with nausea or vomiting, visual disturbances such as scintillating or zigzag light waves. The person may experience a visual aura before an attack actually happens. This is usually called a classical migraine. There are other types of migraine:

- **Ophthalmoplegic migraine** – this causes painful, red and watery eyes.

- **Neuropathic migraine** – this causes one-sided paralysis and weakness of the face and body.

- **Abdominal migraine** – this can affect children with recurring attacks of abdominal pain, sometimes accompanied by nausea and vomiting.

Migraines can be treated with simple analgesia or more specialised anti-migraine medication.

Motor neurone disease

This is a progressive, degenerative disease of the motor neurones of the nervous system. It tends to occur in middle age and causes muscle weakness and wasting.

Multiple sclerosis

This is a disease of the central nervous system in which the myelin (fatty) sheath covering the nerve fibres is destroyed and various functions become impaired, including movement and sensations. Multiple sclerosis is characterised by relapses and remissions. It can present with blindness or reduced vision and can lead to severe disability within a short period. It can also cause incontinence, loss of balance, tremor and speech problems. Depression and mania can occur.

Neuralgia

Neuralgia presents as attacks of pain along the entire course or branch of a peripheral sensory nerve. A common example is trigeminal neuralgia affecting the trigeminal nerve in the face.

Neuritis

This is an inflammation/disease of a single or several nerves with different causes, such as infection, injury or poison. It causes pain along the length of the nerve and/or loss of the use of structures supplied by the nerve.

Parkinson's disease

This disease is caused by damage to the grey matter of the brain known as basal ganglia. It causes involuntary tremors of limbs, with stiffness, rigidity and a shuffling gait. The face lacks expression and movements are slow. People may suffer from depression, confusion and anxiety.

Sciatica

This is lower back pain which can affect the buttock and thigh. On occasions it radiates to the leg and foot. In severe cases it can cause numbness and weakness of the lower limb. It can result from prolapse of the discs between the spinal vertebrae, tumour or blood clot (thrombosis). Diabetes or heavy alcohol intake can also produce symptoms of sciatica. This condition tends to recur and may require strong analgesia or surgery in severe cases.

Stress

Stress can be defined as any factor that affects physical or emotional well-being. Signs of stress affecting the nervous system include anxiety, depression, irritability, headaches, back pain and excessive tiredness.

Interrelationships with other systems

The nervous system links to the following body systems:

Cells and tissues

Nervous tissue is a specialised type of tissue which can pick up and transmit electrical signals by converting stimuli into nerve impulses.

Skin

The skin is a highly sensitive organ and has many sensory nerve endings which respond to touch, temperature and pressure.

Skeletal

The skeleton provides protection for the spinal cord and the brain.

Muscular

The brain sends impulses to muscles via motor nerves in order to effect movement.

Circulatory

Blood transports vital oxygen to the nerve cells. The medulla oblongata in the brain is the control centre for the heart. The sympathetic nervous system prepares the body for activity by increasing the heart rate. The parasympathetic nervous system encourages the resting heart rate.

Respiratory

Oxygen inhaled into the body is carried to the nerve cells to enable them to function properly. Without oxygen nerve cells become damaged and die, causing irreversible damage. The sympathetic nervous system prepares the body for activity by increasing the respiration rate. The parasympathetic nervous system encourages the resting respiratory rate.

Endocrine

The endocrine system works closely with the nervous system in order to maintain homeostasis in the body.

Digestive

The nervous system influences the actions of the digestive system. The sympathetic nervous system effects include increased conversion of glycogen to glucose by the liver and decreased secretion of saliva. The parasympathetic nervous system effects include increased gastrointestinal activity and stimulated salivation.

Key words associated with the nervous system

central nervous system	spinal nerves	meninges
brain	cervical	cerebrum
spinal cord	thoracic	thalamus
peripheral nervous system	lumbar	hypothalamus
somatic nervous system	sacral	pineal gland
cranial nerves	coccygeal	cerebellum
olfactory	plexuses	brain stem
optic	autonomic nervous system	pons
oculomotor	neurone	medulla oblongata
trochlear	neuroglia	spinal cord
trigeminal	cell body	reflex action
abducens	dendrites	sense organs
facial	axon	nose
vestibulocochlear	sensory neurone	eyes
glossopharyngeal	motor neurone	ears
vagus	mixed neurone	tongue
accessory	synapse	skin
hypoglossal	chemical transmitter substance	

Summary of the nervous system

- The nervous system helps regulate homeostasis and integrate all body activities by sensing changes, interpreting them and reacting to them.

- The **central nervous system** (CNS) consists of the brain and the spinal cord.

- The **peripheral nervous system** (PNS) consists of the somatic nervous system, made up of the cranial and spinal nerves and the autonomic (involuntary) nervous system.

- There are two types of nervous tissue.

  - **Neurone** – this is a functional unit of the nervous system. The neurone is designed to receive stimuli and conduct impulses.

  - **Neuroglia** – this is a specialised type of connective tissue that supports, nourishes and protects neurones.

- Neurones have two major properties – excitability and conductibility.

- Most **nerve cells**, or neurones, consist of a cell body, many dendrites and usually a single axon.

- There are three main types of neurones – sensory, motor and mixed.

- **Sensory neurones** conduct impulses from receptors to the CNS.

- **Motor neurones** conduct impulses to effectors (muscles).

- **Mixed neurones** conduct impulses to other neurones.

- The junction where nerve impulses are transmitted from one neurone to another is called a **synapse**.

- **Impulses** are relayed from one neurone to another by a chemical transmitter substance which is released by the neurone to carry impulses across the synapse to stimulate the next neurone.

- The **central nervous system** (brain and spinal cord) is covered by a special protective type of connective tissue in three layers called the **meninges**.

- The parts of the brain include the **cerebrum**, **thalamus**, **hypothalamus**, **pituitary gland**, **pineal gland**, **cerebellum** and the **brain stem**.

- The **cerebrum** is the largest part of the brain and is concerned with all forms of conscious activity. It has sensory areas which control vision, touch, hearing, taste and smell, and also motor areas which control voluntary movements and association areas which control reasoning, memory and emotions.

- The **thalamus** is a relay and interpretation centre for all sensory impulses, except olfaction.

- The **hypothalamus** controls hunger, thirst, temperature regulation, anger, aggression, hormones, sexual behaviour, sleep patterns and consciousness.

- The **pineal gland** is involved in the regulation of circadian rhythms and is thought to influence mood.

- The **cerebellum** is concerned with the coordination of skeletal muscles, muscle tone and balance.

- The **brain stem** contains the midbrain, pons and medulla oblongata.

- The **midbrain** contains certain visual and auditory reflexes that coordinate head and eye movements with things seen and heard. The **pons** relays messages from the cerebral cortex to the spinal cord and helps regulate breathing.

- The **medulla oblongata** contains control centres for the heart, lungs and intestines.

- The **spinal cord** is an extension of the brain stem and its function is to relay impulses to and from the brain.

- A **reflex action** is a rapid and automatic response to a stimulus without any conscious action of the brain.

- The **peripheral nervous system** contains all the nerves outside of the central nervous system and can be subdivided into the **somatic nervous system** and the **autonomic nervous system**.

- The **somatic nervous system** contains **31 pairs of spinal nerves** (nerves originating from the spinal cord) and **12 pairs of cranial nerves** (nerves originating from the brain).

- The 31 pairs of spinal nerves are **8 cervical, 12 thoracic, 5 lumbar, 5 sacral and 1 coccygeal.**

- Each **spinal nerve** is divided into several branches, forming a network of nerves or **plexuses** which supply different parts of the body.

- The 12 pairs of cranial nerves connect directly to the brain. They are **olfactory, optic, oculomotor, trochlear, trigeminal, abducens, facial, vestibulocochlear, glossopharyngeal, vagus, accessory and hypoglossal.**

- **The autonomic nervous system** is the part of the nervous system that controls the automatic body activities of smooth and cardiac muscle and the activities of glands. It is divided into the **sympathetic** and **parasympathetic** divisions.

- The activity of the **sympathetic** system is to prepare the body for expending energy and dealing with emergency situations.

- The **parasympathetic** system balances the action of the sympathetic division by working to conserve energy and create the conditions needed for rest and sleep. It slows down the body processes, except digestion and the functions of the genito-urinary system.

- The **sense organs** include the **nose** (olfaction), **tongue** (taste), **eyes** (sight), **ears** (hearing) and **skin** (touch).

Multiple-choice questions

1. **The nervous system is divided into two major divisions:**
   a central nervous system and peripheral nervous system
   b central nervous system and autonomic nervous system
   c the brain and the spinal cord
   d the peripheral nervous system and the brain

2. **The components of the central nervous system include:**
   a the spinal cord and cranial nerves
   b the sympathetic and parasympathetic nervous systems
   c the brain and spinal cord
   d the cranial and spinal nerves

3. **The three basic parts of a neurone are:**
   a cell body, sensory and afferent nerves
   b cell body, nucleus and axon
   c cell body, axon and dendrites
   d cell body, motor and efferent nerves

4. **A sensory nerve is responsible for sending messages:**
   a from the brain and spinal cord
   b to and from the brain and spinal cord
   c to the brain and spinal cord
   d none of the above

5. **The part of the brain that houses the thalamus and hypothalamus is the:**
   a cerebrum
   b brain stem
   c cerebellum
   d medulla oblongata

6. **The region of the brain concerned with the coordination of skeletal muscle is the:**
   a cerebellum
   b pons
   c midbrain
   d cerebrum

7. **The part of the brain concerned with all forms of conscious activity is the:**
   a cerebrum
   b thalamus
   c hypothalamus
   d medulla oblongata

8. **The connective tissue membranes that envelop the central nervous system are:**
   a cerebrospinal membranes
   b meninges
   c myelin sheaths
   d synapses

9. **The part of the brain that contains vital control centres for the heart, lungs and intestines is the:**
   a hypothalamus
   b midbrain
   c medulla oblongata
   d cerebellum

10. **The junction where nerve impulses are transmitted from one neurone to another is a:**
    a neurotransmitter
    b synapse
    c dendrite
    d axon

11. The point where the nerve supply enters the muscle is the:

a motor impulse
b motor transmitter
c motor point
d muscle fibre

12. The spinal cord is an extension of which part of the brain?

a pons
b medulla oblongata
c midbrain
d brain stem

13. Which of the following statements is true?

a sensory neurones receive stimuli from sensory organs and receptors and transmit the impulses to the spinal cord and brain
b motor neurones conduct impulses to the brain and spinal cord and to the muscles and glands
c association neurones link motor neurones together
d axons receive and transmit stimuli away from the cell body

14. Which of the following is not one of the cranial nerves?

a trigeminal
b facial
c cervical
d optic

15. The effects of the parasympathetic nervous system are:

a resting heart rate
b increased gastrointestinal activity
c pupil constriction
d all of the above

the endocrine system

Introduction

The endocrine system comprises a series of internal secretions called hormones which help to regulate body processes by providing a constant internal environment. Hormones are chemical messengers and act as catalysts in that they affect the physiological activities of other cells in the body. The endocrine system works closely with the nervous system. Nerves enable the body to respond rapidly to stimuli, whereas the endocrine system causes slower and longer-lasting effects.

Objectives

By the end of this chapter you will be able to recall and understand the following knowledge:

- the functions of the endocrine system
- the definition of a hormone
- the location of the main endocrine glands of the body
- the principal hormone secretions from the main endocrine glands and their effects on the body
- the natural glandular changes that occur in the body such as puberty, menstruation, pregnancy and menopause
- the interrelationships between the endocrine and other body systems
- disorders of the endocrine system.

Functions of the endocrine system

The functions of the endocrine system are:

- producing and secreting hormones which regulate body activities such as growth, development and metabolism
- maintaining the body during times of stress
- contributing to the reproductive process.

What is a hormone?

A hormone is a chemical messenger or regulator, secreted by an endocrine gland which reaches its destination by the bloodstream and has the power of influencing the activity of other organs. Some hormones have a slow action over a period of years, such as the growth hormone from the anterior pituitary, while others have a quick action, such as adrenaline from the adrenal medulla. Hormones, therefore, regulate and coordinate various functions in the body.

The endocrine glands are ductless glands, as the hormones they secrete pass directly into the bloodstream to influence the activity of another organ or gland. The main endocrine glands are as follows:

- pituitary gland
- thyroid gland
- parathyroid glands
- adrenal glands
- islets of Langerhans
- ovaries in the female
- testes in the male.

Pituitary gland

This is a lobed structure, attached by a stalk to the hypothalamus of the brain. For many years the pituitary gland has been referred to as the 'master' endocrine gland because it secretes several hormones that control other endocrine glands. However, the pituitary itself has a master – the hypothalamus.

The hypothalamus is a small region of the brain that is the major integrating link between the nervous and endocrine systems. Hormones of the pituitary are controlled by releasing or inhibiting hormones produced by the hypothalamus.

The hypothalamus initiates the process by producing its own set of hormones (releasing or inhibiting hormones) as a result of stimulation in the brain. This has a cascading effect on the pituitary, which in turn produces its own hormones that stimulate other glands. An example is thyrotrophin-releasing hormone from the hypothalamus that promotes the pituitary to secrete thyroid-stimulating hormone which controls the growth and activity of the thyroid gland.

The pituitary gland consists of two main parts – an anterior and a posterior lobe.

Anterior lobe

The principal hormones secreted by the anterior lobe of the pituitary are as follows:

- **Growth hormone** – this controls the growth of long bones and muscles.

- **Thyroid-stimulating hormone (TSH)** – this controls the growth and activity of the thyroid gland.

- **Adrenocorticotrophic hormone (ACTH)** – this stimulates and controls the growth and hormonal output of the adrenal cortex.

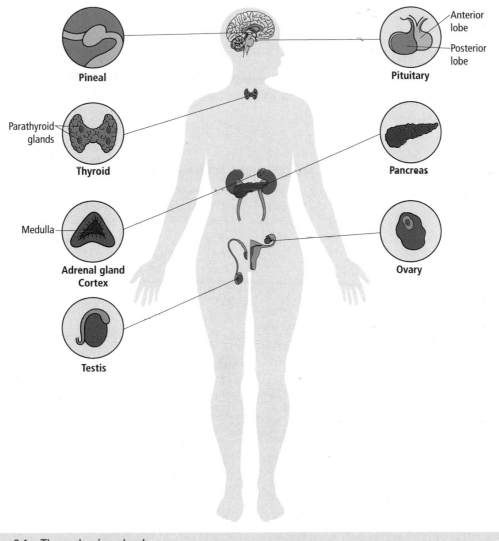

Figure 9.1   The endocrine glands

■ **Gonadotrophic hormones** – these control the development and growth of the ovaries and testes.

The gonads or sex hormones include:

■ **Follicle-stimulating hormone** – in women this stimulates the development of the Graafian follicle in the ovary which secretes the hormone oestrogen. In men it stimulates the testes to produce sperm.

■ **Luteinizing hormone** – in women this helps to prepare the uterus for the fertilised ovum. In men it acts on the testes to produce testosterone.

■ **Prolactin** – this stimulates the secretion of milk from the breasts following birth.

■ **Melanocyte-stimulating hormone (MSH)** – this stimulates the production of melanin in the basal cell layer of the skin.

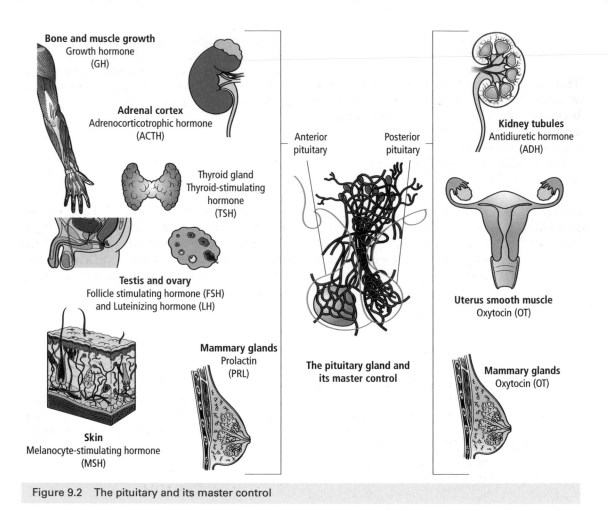

Bone and muscle growth
Growth hormone
(GH)

Adrenal cortex
Adrenocorticotrophic hormone
(ACTH)

Thyroid gland
Thyroid-stimulating
hormone
(TSH)

Testis and ovary
Follicle stimulating hormone (FSH)
and Luteinizing hormone (LH)

Mammary glands
Prolactin
(PRL)

Skin
Melanocyte-stimulating hormone
(MSH)

Anterior
pituitary

Posterior
pituitary

**The pituitary gland and
its master control**

Kidney tubules
Antidiuretic hormone
(ADH)

Uterus smooth muscle
Oxytocin (OT)

Mammary glands
Oxytocin (OT)

Figure 9.2   The pituitary and its master control

Key note

Endocrine glands in the body have a feedback mechanism which is coordinated by the pituitary gland. This gland is influenced by the hypothalamus and will increase its output of releasing factors if other glands start to fail, or decrease its output if the level of the hormone in the bloodstream starts to rise.

Posterior lobe

The posterior lobe of the pituitary secretes two hormones which are manufactured in the hypothalamus but are stored in the posterior lobe:

- **Antidiuretic hormone** (ADH) – this increases water reabsorption in the renal tubules of the kidneys.

- **Oxytocin** – this stimulates the uterus during labour and stimulates the breasts to produce milk.

Pineal gland

This is a pea-sized mass of nerve tissue attached by a stalk in the central part of the brain. It is located deep between the cerebral hemispheres where it is attached to the upper portion of the thalamus.

The pineal gland functions as a gland and secretes a hormone called melatonin which synthesises from serotonin. This gland is involved in the regulation of circadian rhythms, patterns of repeated activity that are associated with the environmental cycles of day and night, such as sleep and wake rhythms. It is also thought to influence moods.

Thyroid gland

The thyroid gland is found in the neck, situated on either side of the trachea, and is controlled by the anterior lobe of the pituitary. The principal secretions of the thyroid gland are **triodothyronine (T3)** and **thyroxine (T4)** which both regulate growth and development and also influence mental, physical and metabolic activities. The thyroid gland also secretes the hormone **calcitonin** which controls the level of calcium in the blood.

The functions of the thyroid gland are as follows:

- controls the metabolic rate by stimulating metabolism
- influences growth and cell division
- influences mental development
- is responsible for the maintenance of healthy skin and hair
- stores the mineral iodine which it needs to manufacture thyroxin
- stimulates the involuntary nervous system and controls irritability.

The thyroid gland is controlled by a feedback mechanism. It will increase to meet the demand for more thyroid hormones at various times such as during the menstrual cycle, pregnancy and puberty.

Parathyroid glands

These are four small glands situated on the posterior of the thyroid gland. Their principal secretion is the hormone parathormone which helps to regulate calcium metabolism by controlling the amount of calcium in blood and bones.

Adrenal glands

These are two triangular-shaped glands which lie on top of each kidney. They consist of two parts – an outer cortex and an inner medulla.

Adrenal cortex

The principal hormones secreted by the adrenal cortex are as follows:

- **Glucocorticoids** (cortisone and hydrocortisone) – these hormones influence the metabolism of protein and carbohydrates and utilise fats. They are important in maintaining the level of glucose in the blood so that blood glucose levels are increased at times of stress.
- **Mineral corticoids** (aldosterone) – this hormone acts on the kidney tubules, retaining salts in the body, excreting excess potassium and maintaining the water and electrolyte balance.

■ **Sex corticoids** – these include testosterone, oestrogen and progesterone. These hormones control the development of the secondary sex characteristics and the function of the reproductive organs.

Key note

When the ovaries and testes mature, they produce the sex hormones themselves, therefore the production of sex corticoids in the adrenal cortex is important up to puberty.

Adrenal medulla

The principal hormones secreted by the adrenal medulla are adrenaline and nor-adrenaline. They are under the control of the sympathetic nervous system and are released at times of stress. The responses of these hormones are fast due to the fact that they are governed by nervous control.

The effects of these stress hormones are similar, although adrenaline has a primary influence on the heart, causing an increase in heart rate, whereas noradrenaline has a greater effect on peripheral vasoconstriction which raises blood pressure.

A summary of the effects of adrenaline are as follows:

■ dilates the arteries, increasing blood circulation and the heart rate

■ dilates the bronchial tubes, increasing oxygen intake and the rate and depth of breathing

■ raises the metabolic rate

■ constricts the blood vessels to the skin and intestines, diverting blood from these regions to the muscles and brain to effect action.

The effects of noradrenaline are similar to those of adrenaline and include:

■ vasoconstriction of small blood vessels leading to an increase in blood pressure

■ increase in the rate and depth of breathing

■ relaxation of the smooth muscle of the intestinal wall.

Key note

The effects described above are those felt when the body is under stress, such as a pounding heart, increased ventilation rate, dry mouth and butterflies in the stomach. Levels of stress hormones are broken down slowly so that effects on the sympathetic nervous system are long-lasting. Over the long term, if levels of these hormones remain elevated, they perpetuate factors for stress-related disorders.

Pancreas

The pancreas is known as a dual organ as it has an endocrine and an exocrine function. The exocrine or external secretion is the secretion of pancreatic juice to assist with digestion. The endocrine or internal secretion is the hormone **insulin**, secreted by the islets of Langerhans cells in the pancreas. Insulin lowers the level of sugar in the blood by helping the body cells to take it up and use it or store it as glycogen.

Sex glands

Testes

The testes are situated in the groin in a sac called the scrotum. They have two functions:

- the secretion of the hormone testosterone which controls the development of the secondary sex characteristics in the male at puberty (influenced by the luteinising hormone)

- the production of sperm (influenced by the follicle-stimulating hormone from the anterior pituitary).

Ovaries

The ovaries are situated in the lower abdomen below the kidneys. The two ovaries are the sex glands in the female, and each is attached to the upper part of the uterus by broad ligaments. The ovaries have two distinct functions:

- production of ova at ovulation

- production of the female sex hormones oestrogen and progesterone.

Oestrogen is concerned with the development and maintenance of the reproductive system and the development of the secondary sex characteristics. Progesterone is produced by the ovaries after ovulation. It helps to prepare the uterus for the implantation of the fertilised ovum, develops the placenta and prepares the breasts for milk secretion.

The ovaries also secrete the following hormones in addition to oestrogen and progesterone:

- **Inhibin** – this hormone inhibits the secretion of the follicle-stimulating hormone (FSH) towards the end of the menstrual cycle.

- **Relaxin** – this hormone dilates the cervix and assists the pelvis in widening during childbirth.

Natural glandular changes

Puberty

This is the time at which the onset of sexual maturity occurs and the reproductive organs become functional. Changes in both sexes occur with the appearance of the secondary sexual characteristics, such as the deepening of the voice in boys and the growth of breasts in girls. These changes are brought about by an increase in sex hormone activity, due to stimulation of the ovaries and testes by the pituitary gonadotrophic hormones.

The average age for girls to reach puberty is between 10 and 14, although it can occur as early as 8 or 9 years of age. In boys, the average age is 13 to 16.

In girls the ovaries are stimulated by the gonadotrophic hormones – the follicle-stimulating hormone (FSH) and luteinising hormone (LH). The effects of puberty in girls include:

- the onset of ovulation and the menstrual cycle

- the female reproductive organs becoming functional

- the growth of pubic and axillary hair

- development of breast tissue

- increase in the amount of subcutaneous fat.

In boys the same gonadotrophic hormones (FSH and LH) stimulate the testes to produce testosterone. The effects of puberty in boys include:

- voice breaking and larynx enlarging

- the growth of muscle and bone

- noticeable height increase

- the development of sexual organs

- the growth of pubic, facial, axillary, abdominal and chest hair

- the onset of sperm production.

The menstrual cycle

Starting at puberty, the female reproductive system undergoes a regular sequence of monthly events, known as the menstrual cycle. The ovaries undergo cyclical changes in which a certain number of ovarian follicles develop. When one ovum completes the development process, it is released into one of the fallopian tubes. If fertilisation does not occur, the developed ovum disintegrates and a new cycle begins.

The menstrual cycle lasts approximately 28 days, although it can be longer or shorter than this. There are three stages of the menstrual cycle:

- **proliferative** (first) phase – days 7 to 14 of the cycle

- **secretory** (second) phase – days 14 to 28 of the cycle

- **menstrual** (third) phase – days 1 to 7 of the cycle.

Proliferative phase

At the beginning of the cycle an ovum develops within an ovarian follicle in the ovary. This is in response to a hormone released by the anterior lobe of the pituitary gland called the follicle-stimulating hormone (FSH), which stimulates the follicles of the ovaries to produce the hormone oestrogen.

Oestrogen stimulates the endometrium to promote the growth of new blood vessels and mucus-producing cells. When mature, it bursts from the follicle and travels along the fallopian tube to the uterus. This occurs about 14 days after the start of the cycle and is known as ovulation.

Secretory phase

A temporary endocrine gland, the corpus luteum, develops in the ruptured follicle in response to stimulation from the luteinising hormone (LH) secreted by the anterior lobe of the pituitary gland. The corpus luteum secretes the hormone progesterone which, together with oestrogen, causes the lining of the uterus (endometrium) to become thicker and richly supplied with blood in preparation for pregnancy.

After ovulation, the ovum can only be fertilised during the next 8 to 24 hours. If fertilisation does occur, the fertilised ovum becomes attached to the endometrium and the corpus luteum continues to secrete progesterone. Pregnancy then begins. The corpus luteum continues to secrete progesterone until the fourth month of pregnancy, by which time the placenta has taken over this function.

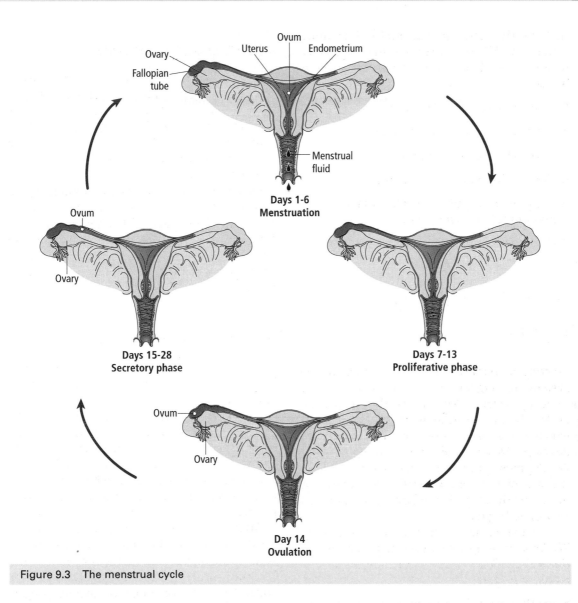

Figure 9.3   The menstrual cycle

Menstrual phase

If the ovum is not fertilised, the cycle continues and the corpus luteum shrinks and the endometrium is shed. This is called menstruation. Over a period of about five days, the muscles of the wall of the uterus contract to expel the unfertilised egg, pieces of endometrial tissue and some tissue fluid.

As soon as levels of progesterone drop, due to the breakdown of the endometrium and the corpus luteum, the pituitary gland starts producing progesterone again and hence stimulates the ovaries to produce another follicle and a new ovum. The cycle then begins again.

Pregnancy

Pregnancy takes approximately nine calendar months and is divided into three trimesters:

- The first trimester is the most important to the developing baby and is a time of radical hormonal changes. During this phase all the body systems develop.

- The second trimester consists of rapid foetal growth and the completion of systemic development. Blood volume for the mother increases as additional workload is placed on all physiological functions. Cardiac output, breathing rate and urine production increase in response to foetal demands. The uterus enlarges greatly during pregnancy, along with the size of the breasts. Appetite increases in response to the foetal need for increasing amounts of nutrients.

- The third trimester is mostly a weight-gaining and maturing process, preparing the baby for life outside the womb. Posture changes are evident at this stage as the mother gains more weight and internal organs are compressed. The body's connective tissue structure alters by softening, to allow for the expansion needed for the birth.

Hormonal changes that occur during pregnancy

During a typical menstrual cycle, the corpus luteum degenerates about two weeks after ovulation. Consequently, the levels of oestrogen and progesterone decline rapidly and the lining of the uterus (endometrium) is not maintained and is cast off as menstrual flow.

If this occurs after implantation the embryo becomes spontaneously aborted (miscarries). The mechanism that usually prevents this occurring involves a hormone called human chronic gonadotrophic hormone (HCG), which is secreted by a layer of embryonic cells that surround the developing embryo. HCG causes the corpus luteum to be maintained in order to establish the pregnancy.

The maintenance of the corpus luteum is important for the first three months, after which the placenta is usually well developed and is able to secrete sufficient oestrogen and progesterone.

The secretion of the hormones oestrogen and progesterone is important during pregnancy as they:

- maintain the uterine wall

- inhibit the secretion of the gonadotrophic hormones FSH and LH

- stimulate development of the mammary glands

- inhibit uterine contractions until birth

- cause enlargement of the reproductive organs.

The ovaries and placenta produce inhibin which inhibits the secretion of the follicle-stimulating hormone (FSH) from the anterior lobe of the pituitary, thus preventing the development of ova during pregnancy.

At the end of the gestation period the levels of progesterone fall. Labour cannot begin until the levels of progesterone fall as progesterone inhibits uterine contractions. Oxytocin, secreted by the posterior lobe of the pituitary, stimulates uterine contractions and the ovaries and placenta secrete relaxin which helps to dilate the cervix and relaxes the ligaments and joints to assist labour.

Key note

Mood, sleep and energy levels are all affected during pregnancy due to the hormonal changes that occur during this period. Some women may report extreme tiredness at the start of their pregnancy and experience a surge of energy towards the end. Sleep patterns may be affected due to the activity of the growing foetus and hormone levels may promote tearfulness and emotional disturbances.

Menopause

After puberty the menstrual cycle normally continues to occur at regular intervals into approximately the late forties or early fifties (most commonly between 45 and 55). At this time there are marked changes in which the cycle becomes increasingly irregular until the cycle ceases altogether. This period in a woman's reproductive life is called the menopause (female climacteric).

During the menopause there is a change in the balance of the sex hormones. The ovaries cease responding to the follicle-stimulating hormone (FSH) and this decline in function results in lower levels of oestrogen and progesterone secretion. As a result of reduced oestrogen concentration and lack of progesterone, the female's secondary sexual characteristics undergo varying degrees of change, which may include a decrease in the size of the vagina, uterus and uterine tubes, as well as atrophy of the breasts.

Other changes that occur commonly in response to low oestrogen concentration include an increased loss of bone matrix, increasing the risk of osteoporosis, thinning of the skin and dryness of the mucous membrane lining the vagina.

Some women of menopausal age experience unpleasant vasomotor symptoms, including sensations of heat in the face, neck and upper body, known as hot flushes. Menopausal women may also experience varying degrees of headache, backache and fatigue, as well as emotional disturbances.

Key note

Menopausal women are often treated with hormone replacement therapy (HRT) in order to alleviate some of the unpleasant side effects. HRT usually involves administering oestrogens, along with progesterone.

Disorders of the endocrine system

Addison's disease

Undersecretion of corticosteroid hormones is responsible for the condition known as Addison's disease. Symptoms include loss of appetite, weight loss, brown pigmentation around joints, low blood sugar, low blood pressure, tiredness and muscular weakness. This disease is treatable by replacement hormone therapy.

Cretinism

Hyposecretion of thyroxine leads to cretinism in children which is a congenital deficiency causing impaired mentality, small stature, coarsening of the skin and hair and deposition of fat on the body.

Cushing's syndrome

Hypersecretion of the glucocorticoids can lead to a condition known as Cushing's syndrome. This condition results from an excess amount of corticosteroid hormones in the body. Symptoms include weight gain, reddening of the face and neck, excess growth of facial and body hair, raised blood pressure, loss of mineral from bone and sometimes mental disturbances.

Diabetes insipidus

Hyposecretion of the antidiuretic hormone by the posterior lobe of the pituitary leads to the disease diabetes insipidus. Symptoms include dehydration, increased thirst and increased output of urine.

Diabetes mellitus

Hyposecretion can lead to a condition called diabetes mellitus. This condition is due to a deficiency or absence of insulin. The symptoms associated with diabetes include an increased thirst, increased output of urine, weight loss, thin skin with impaired healing capacity, increased tendency to develop minor skin infections and decreased pain threshold when insulin levels are low. There are two types of diabetes mellitus:

- **Insulin-dependent diabetes** (early onset) – this occurs mainly in children and young adults and the onset is usually sudden. The deficiency or absence of insulin is due to the destruction of the islet cells in the pancreas. The causes are unknown but there is a familial tendency, suggesting genetic involvement.

- **Non-insulin-dependent diabetes** (late onset): this type of diabetes occurs later in life and its causes are unknown. Insulin secretion may be below or above normal. Deficiency of glucose inside the body cells may occur where there is hyperglycaemia and a high insulin level. This may be due to changes in cell walls which block the insulin-assisted movement of glucose into cells. This type of diabetes can be controlled by diet alone, or diet and oral drugs.

Dwarfism

Hyposecretion of the growth hormone during childhood leads to stunted growth, a condition known as dwarfism.

Gigantism

Hypersecretion of the growth hormone secreted by the anterior pituitary leads to gigantism in children, a disease marked by the rapid growth of the body to extremely large proportions (seven to eight feet). If the overproduction occurs in adulthood, then there is an abnormal enlargement of the hands and feet and coarsening of the facial features, due to the continued growth of tissues. This condition is known as acromegaly.

Gynaecomastia

Hypersecretion of oestrogen and progesterone in the male can lead to muscle atrophy and breast development.

Hirsutism

This is hair growth in the male sexual pattern due to hypersecretion of the hormone testosterone in women and an overproduction of androgens.

Hypoglycaemia

Hypoglycaemia (low blood sugar level) can lead to muscular weakness and incoordination, mental confusion and sweating. If severe it may lead to a hypoglycaemic coma.

Myxoedema

Hyposecretion of thyroxine in an adult leads to myxoedema which is characterised by the slowing down of physical and mental activity, resulting in lethargy, hair becoming brittle, skin becoming coarse and dry and a slow metabolism.

Polycystic ovary syndrome

Hyposecretion of the hormones oestrogen and progesterone in the female can lead to polycystic ovary syndrome which is characterised by cysts on the ovaries, cessation of periods, obesity, atrophy of the breasts, hirsutism and sterility.

Seasonal affective disorder (SAD)

Hyposecretion of the hormone melatonin is thought to be associated with the condition seasonal affective disorder (SAD). Symptoms include depression (typically with the onset of winter), a general slowing down of mind and body and excessive sleeping and overeating.

Stress

Stress can be defined as any factor which affects physical or emotional health. Effects of short-term physical stress are associated with the hormone adrenaline and include an increased heart beat, rapid breathing, increased sweating, tense muscles, dry mouth, frequency of urination and a feeling of nausea. Stress can become negative stress when excess adrenaline is left in the bloodstream, following a short-term stress signal. Examples of possible symptoms of excessive stress on the endocrine system include amenorrhea (absence or stopping of periods), loss of libido and infertility.

Tetany

When there is a deficiency of calcium in the blood a condition known as tetany occurs which is characterised by muscular twitchings particular to the hands and feet. These symptoms are quickly relieved by administering calcium.

Thyrotoxicosis

Hypersecretion of thyroxine leads to a condition known as thyrotoxicosis or Graves' disease. Thyrotoxicosis results in an increased metabolic rate, weight loss, sweating, restlessness, increased appetite, sensitivity to heat, raised temperature, frequent bowel action, anxiety and nervousness. When the thyroid gland produces and secretes an excessive amount of thyroxine, it may produce a goitre (an enlargement of the thyroid gland).

Virilism

Hypersecretion of the hormone testosterone in women can lead to virilism (masculinisation), causing an overproduction of androgens.

Interrelationships with other systems

The endocrine system links to the following body systems:

Cells and tissues

Meiosis is the form of cell division involving the formation of sperm in the male and ova in the female.

Skin

MSH (melanocyte-stimulating hormone) produced by the central lobe of the pituitary gland stimulates the production of melanin in the basal cell layer (stratum germinativum) of the skin.

Skeletal

The hormones calcitonin from the thyroid gland and parathormone from the parathyroid glands help to maintain the calcium levels in the blood for strength and flexibility.

Muscular

Muscles receive additional blood flow in response to the secretion of the hormone adrenaline at times of stress.

Circulatory

Hormones are secreted and carried in the bloodstream to their target organs.

Respiratory
The adrenal glands increase the breathing rate in times of stress to provide more oxygen as fuel for the muscles.

Nervous
The endocrine system works closely with the nervous system in order to maintain homeostasis in the body. The endocrine system is linked to the nervous system by the hypothalamus and the pituitary gland.

Digestive
The production of insulin and glucagon in the pancreas helps to regulate blood sugar levels.

Urinary
The antidiuretic hormone (ADH) helps to regulate fluid balance in the body.

Key words associated with the endocrine system

ductless glands	dwarfism	pancreas
hormone	diabetes insipidus	insulin
pituitary	melanocyte-stimulating	hypoglycaemia
thyroid	hormone (MSH)	diabetes mellitus
parathyroids	pineal gland	testosterone
adrenals	melatonin	oestrogen
islets of Langerhans	triodothyronine (T3)	progesterone
ovaries	thyroxine (T4)	virilism
testes	calcitonin	hirsutism
growth hormone	thyrotoxicosis/Graves'	amenorrhea
thyroid-stimulating hormone	disease	gynaecomastia
(TSH)	cretinism	polycystic ovary syndrome
adrenocorticotrophic	myxoedema	puberty
hormone (ACTH)	tetany	menstrual cycle
gonadotrophic hormones	glucocorticoids	ovarian follicles
(FSH and LH)	mineral corticoids	ovum
prolactin	sex corticoids	fallopian tubes
antidiuretic hormone (ADH)	Cushing's syndrome	fertilisation
oxytocin	Addison's disease	pregnancy
gigantism	adrenaline	trimester
acromegaly	noradrenaline	menopause

Summary of the endocrine system

- The endocrine system consists of ductless glands that secrete hormones into the bloodstream.

- Endocrine glands are concerned with the regulation of metabolic processes.

- A **hormone** is a chemical regulator secreted by an endocrine gland into the bloodstream and has the power to influence the activity of other organs.

- The main endocrine glands are the **pituitary** (attached to base of brain), **thyroid** (neck), **parathyroids** (posterior to the thyroid glands), **adrenals** (top of kidneys), **islets of Langerhans** (in the pancreas), **ovaries** (in the female) and **testes** (in the male).

- The principal hormones secreted by the **anterior lobe of the pituitary** include the **growth hormone, thyroid-stimulating hormone (TSH), adrenocorticotrophic hormone (ACTH), gonadotrophic hormones (FSH and LH), prolactin** and **melanocyte-stimulating hormone (MSH)**.

- The **growth hormone** controls the growth of long bone and muscle.

- **Thyroid-stimulating hormone (TSH)** controls the growth and activity of the **thyroid gland.**

- **Adrenocorticotrophic hormone (ACTH)** controls the growth and hormonal output of the **adrenal cortex.**

- **Gonadotrophic hormones (FSH and LH)** control the development and growth of the **ovaries** and **testes.**

- **Prolactin** stimulates the secretion of milk from the breasts following birth.

- **Melanocyte-stimulating hormone (MSH)** stimulates the production of melanin in the basal cell layer of the skin.

- **Hypersecretion** of the growth hormone from the pituitary gland can lead to **gigantism** in childhood and **acromegaly** in adulthood.

- **Hyposecretion** of the growth hormone from the pituitary gland during childhood leads to dwarfism.

- The posterior lobe of the pituitary secretes the **antidiuretic hormone (ADH)** and **oxytocin.**

- **Hyposecretion** of the antidiuretic hormone (ADH) by the posterior lobe of the pituitary can lead to **diabetes insipidus.**

- The **pineal gland** is attached by a stalk in the central part of the brain and secretes a hormone called **melatonin** which is thought to regulate circadian rhythms and influence mood.

- The **thyroid gland**'s principal secretions are **triodothyronine (T3)** and **thyroxine (T4)**, which regulate metabolism and influence growth and development.

- The thyroid gland also secretes **calcitonin** which controls the levels of calcium in the blood.

- **Hypersecretion** of the thyroid hormones leads to a condition called **thyrotoxicosis** or Graves' disease.

- **Hyposecretion** of the thyroid hormones leads to cretinism in childhood and **myxoedema** in adulthood.

- The **parathyroid glands** help regulate calcium metabolism.

- **Hypersecretion** of **parahormone** can lead to renal stones, kidney failure and softening of the bones and tumours.

- **Hyposecretion** of **parahormone** can lead to a condition called **tetany.**

- The **adrenal glands** have two parts – an outer **cortex** and an inner **medulla.**

- The principal hormones secreted by the **adrenal cortex** include **glucocorticoids, mineral corticoids** and **sex corticoids**.

- **Glucocorticoids** influence the metabolism of protein, carbohydrates and utilisation of fats.

- **Mineral corticoids** are concerned with maintaining water and electrolyte balance.

- **Sex corticoids** control the development of the secondary sex characteristics and the function of the reproductive organs.

- **Hypersecretion** of the **mineral corticoids** can lead to kidney failure, high blood pressure and an excess of potassium in the blood.

- **Hypersecretion** of the **glucocorticoids** can lead to a condition called **Cushing's syndrome.**

- **Hypersecretion** of the **sex corticoids** can lead to **hirsutism** and **amenorrhea** in the female and muscle atrophy and development of breasts in the male.

- **Hyposecretion** of the corticosteroid hormones can lead to a condition called **Addison's disease.**

- The principal hormones secreted by the adrenal medulla include adrenaline and noradrenaline.

- **Adrenaline** and **noradrenaline** are under the control of the **sympathetic nervous system** and are released at times of stress.

- The **pancreas** is a dual organ as it has dual functions – exocrine and endocrine.

- The exocrine function is the secretion of pancreatic juice to assist with digestion.

- The endocrine function is the secretion of **insulin** from the **islets of Langerhans** cells which helps to regulate blood sugar levels.

- **Hypersecretion** can lead to **hypoglycaemia**.

- **Hyposecretion** can lead to a condition called **diabetes mellitus.**

- The **testes** (in the male) have two functions – the secretion of **testosterone** and the production of sperm.

- The **ovaries** (in the female) have two functions – the production of ova and production of the hormones **oestrogen** and **progesterone.**

- **Hypersecretion** of the hormone **testosterone** in women can lead to **virilism, hirsutism** and **amenorrhea.**

- **Hypersecretion** of **oestrogen** and **progesterone** in the male can lead to **gynaecomastia.**

- **Hyposecretion** of **oestrogen** and **progesterone** in the female can lead to **polycystic ovary syndrome.**

- **Puberty** is a natural glandular change due to stimulation of the ovaries and testes by the pituitary gonadotrophic hormones.

- Starting at **puberty**, the female reproductive system undergoes a regular sequence of monthly events, known as the **menstrual cycle.**

- The **ovaries** undergo cyclical changes, in which a certain number of **ovarian follicles** develop. When one **ovum** completes the development process, it is released into one of the **fallopian tubes**. If **fertilisation** does not occur, the developed ovum disintegrates and a new cycle begins.

- The **menstrual cycle** lasts approximately 28 days, although it can be longer or shorter than this.

- **Pregnancy** takes approximately nine calendar months and is divided into three trimesters.

- During the first **trimester** all the body systems develop.

- The second **trimester** consists of rapid foetal growth and the completion of systemic development.

- The third **trimester** is mostly a weight-gaining and maturing process, preparing the baby for life outside the womb.

- In the **menopause**, the **ovaries** cease responding to the follicle-stimulating hormone (FSH), resulting in lower levels of oestrogen and progesterone secretion.

Multiple-choice questions

1. **The purpose of the endocrine system is to:**
   a   contribute to the reproductive process
   b   produce and secrete hormones to regulate body activities
   c   maintain the body during times of stress
   d   all of the above

2. **Which of the following statements is false?**
   a   a hormone is a chemical messenger which reaches its destination via the bloodstream
   b   endocrine glands are ductless glands
   c   all hormones have a quick action
   d   hormones regulate and coordinate various functions in the body

3. **Which of the following secretes the adrenocorticotrophic (ACTH) hormone?**
   a   posterior lobe of pituitary
   b   anterior lobe of pituitary
   c   adrenal medulla
   d   adrenal cortex

4. **The endocrine gland responsible for secreting the thyroid-stimulating hormone (TSH) is:**
   a   thyroid gland
   b   anterior lobe of pituitary
   c   posterior lobe of pituitary
   d   parathyroid glands

5. **Which of the following hormones stimulates the uterus during labour?**
   a   prolactin
   b   oestrogen
   c   oxytocin
   d   progesterone

6. **Which of the following hormones increases blood circulation and heart rate?**
   a   insulin
   b   noradrenaline
   c   testosterone
   d   adrenaline

7. **The hormone responsible for increasing water reabsorption in the kidney tubules is:**
   a   luteinising hormone (LH)
   b   antidiuretic hormone (ADH)
   c   follicle-stimulating hormone (FSH)
   d   oxytocin

8. **The islets of Langerhans are situated in the:**
   a   liver
   b   pancreas
   c   ovaries
   d   kidneys

9. **Hyposecretion of the thyroid gland in an adult can lead to a condition called:**
   a   cretinism
   b   Addison's disease
   c   myxoedema
   d   Cushing's syndrome

10. **The hormone that is concerned with the development of the placenta is:**
    a   prolactin
    b   progesterone
    c   follicle-stimulating hormone (FSH)
    d   oestrogen

11. **The effects of adrenaline are:**
    a   increased heart rate
    b   increased metabolic rate
    c   increase of oxygen intake
    d   all of the above

12. The glucocorticoids are secreted by the:

a   adrenal medulla

b   adrenal cortex

c   pancreas

d   pineal gland

13. In puberty the ovaries are stimulated by:

a   gonadotrophic hormones from the anterior pituitary

b   growth hormone from the anterior pituitary

c   prolactin from the anterior pituitary

d   adrenaline from the adrenal medulla

14. Hypersecretion of testosterone in women can lead to the condition:

a   amenorrhea

b   gynaecomastia

c   polycystic ovary syndrome

d   menopause

the reproductive system

chapter 10

Introduction

Of all the body's systems the reproductive system is significantly different between the two sexes. These systems are unique in that they are not vital to the survival of an individual, but they are essential to the continuation of the human species.

Objectives

By the end of this chapter you will be able to recall and understand the following knowledge:

- the functions of the reproductive systems
- the structure and functions of the parts of the female reproductive system
- the structure and functions of the parts of the male reproductive system
- the interrelationships between the reproductive and other body systems
- disorders of the reproductive system.

Functions of the reproductive systems

The male and female reproductive systems are specialised in their dual function to produce the sex hormones responsible for the male and female characteristics and for producing the cells required for reproduction.

Female reproductive system

The function of the female reproductive system is the production of sex hormones and ova (egg cells) which, if fertilised, are supported and protected until birth. The female reproductive system consists of the following internal organs lying in the pelvic cavity:

- **ovaries**
- **uterus**
- **fallopian tubes**
- **vagina.**

The external genitalia is known collectively as the vulva and consists of the:

- labia majora and minora which are liplike folds at the entrance of the vagina

- clitoris, which is attached to the symphysis pubis by a suspensory ligament and contains erectile tissue

- hymen, which is a thin layer of mucous membrane

- greater vestibular glands, which lie in the labia majora, one on each side near the vaginal opening. These glands secrete mucus which lubricates the vulva.

The breasts are accessory glands to the female reproductive system.

Ovaries

These are the female sex glands and they lie on the lateral walls of the pelvis. They are almond-shaped organs which are held in place, one on each side of the uterus, by several ligaments. The largest of the ligaments is the broad ligament which holds the ovaries in close proximity to the fallopian tubes.

The ovary contains numerous small masses of cells called ovarian follicles, within which the ova (egg cells) develop. At the time of birth there are about two million immature ova in the ovaries. Many of the ova degenerate and at the time of puberty there are only about 400,000 left.

The immature ova (or oocytes) lie dormant in the ovary until they are stimulated by a sudden surge in the hormone FSH (follicle-stimulating hormone) at the time of puberty. Normally one egg (ovum) ripens and is released each month.

Functions of the ovaries

The ovaries have two distinct functions:

- the production of ova

- the secretion of the female hormones oestrogen and progesterone.

Oestrogen and progesterone regulate the changes in the uterus throughout the menstrual cycle and pregnancy. Oestrogen is responsible for the development of the female sexual characteristics, while progesterone, produced in the second phase of the menstrual cycle, supplements the action of oestrogen by thickening the lining of the uterus ready for the possible implantation of a fertilised egg.

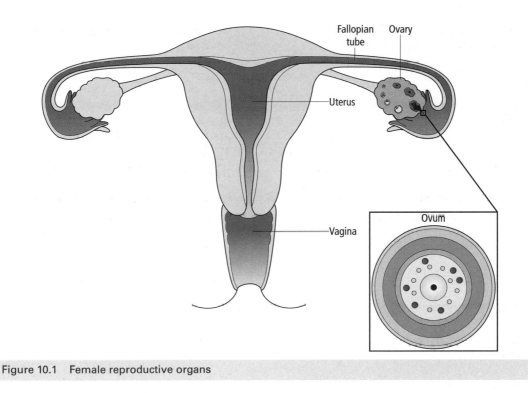

Figure 10.1   Female reproductive organs

Fallopian tubes

The two fallopian tubes are each about 5 cm long and extend from the sides of the uterus, passing upwards and outwards, to end near each ovary. At the end of each fallopian tube are fingerlike projections called fimbriae which encircle the ovaries.

Function of the fallopian tubes

The function of the fallopian tubes is to convey the ovum from the ovary to the uterus. The ovum is swept down the tube by peristaltic muscular contraction, assisted by the lining of ciliated epithelium. Fertilisation of the ovum takes place within the fallopian tubes and it then passes to the uterus.

Uterus

The uterus is a small, hollow, pear-shaped organ situated behind the bladder and in front of the rectum. It has thick, muscular walls and is composed of three layers of tissue:

- **The perimetrium** – this is an outer covering which is part of the peritoneum (a serous membrane in the abdominal cavity). It covers the superior (top) part of the uterus.

- **The myometrium** – this is a middle layer of smooth muscle fibres. This layer forms 90 per cent of the uterine wall and is responsible for the powerful contractions that occur at the time of labour.

■ **The endometrium** – this is a soft, spongy, mucous membrane lining, the surface of which is shed each month during menstruation.

The uterus can be divided into three parts:

■ The fundus is the dome-shaped part of the uterus above the openings of the uterine tubes.

■ The body is the largest and main part of the uterus and leads to the cervix.

■ The cervix of the uterus is a thick, fibrous, muscular structure which opens into the vagina.

Function of the uterus

The uterus is part of the female reproductive tract, which is specialised to receive an ovum and serves as the area in which an embryo grows and develops into a foetus. After puberty the uterus goes through a regular cycle of changes which prepares it to receive, nourish and protect a fertilised ovum.

During pregnancy the walls of the uterus relax to accommodate the growing foetus. If the ovum is not fertilised, the cycle ends with a short period of bleeding in which the endometrium undergoes periodic development and degeneration, known as the menstrual cycle.

Vagina

The vagina is a 10 to 15 cm muscular and elastic tube, lined with moist epithelium, which connects the internal organs of the female reproductive system with the external genitalia. It is made up of vascular and erectile tissue and extends from the cervix (internally) of the uterus above to the vulva (externally) below. During sexual stimulation the erectile tissues become engorged with blood.

Function of the vagina

The function of the vagina is for the reception of the male sperm and to provide a passage-way for menstruation and childbirth. The wall of the vagina is sufficiently elastic to allow for expansion during childbirth. Between the phases of puberty and the menopause, the vagina also provides an acid environment, due to acid-secreting bacteria, in order to help prevent the growth of microbes that may infect the internal organs.

Stages of pregnancy

Pregnancy starts with fertilisation and ends with childbirth.

Fertilisation

This is the fusion of a spermatozoon with an ovum.

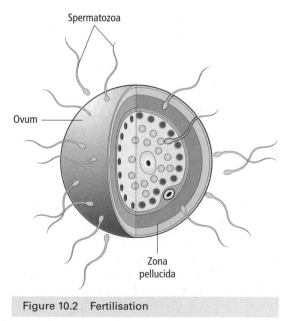

Figure 10.2   Fertilisation

The spermatozoon approaches the ovum and penetrates the inner membrane of the ovum called the zona pellucida. This triggers the ovum's meiotic division following meiosis and makes the zona pellucida impenetrable to other spermatozoon. After the spermatozoon penetrates the ovum, its nucleus is released into the ovum, its tail degenerates and its head enlarges amd fuses with the ovum's nucleus. This fusion provides the fertilised ovum, now called a zygote, with 46 chromosomes.

Pre-embryonic development

The pre-embryonic phase starts with ovum fertilisation and lasts for two weeks. As the zygote passes through the fallopian tube it undergoes a series of mitotic divisions, forming daughter cells initially called blastomeres that each contain the same number of chromosomes.

The first cell division ends about 30 hours after fertilisation. Subsequent divisions occur rapidly. The zygote then develops into a small mass of cells called a morula, which reaches the uterus around the third day after fertilisation. Fluid then masses in the centre of the morula and forms a central cavity which is then called a blastocyst. During the next phase the blastocyst stays within the zona pellucida, unattached to the uterus. The zona pellucida degenerates and by the end of the first week of fertilisation the blastocyst attaches to the endometrium.

Formation of the embryo

By day 24 the blastocyst has formed an amniotic cavity containing an embryo. The developing zygote starts to take on a human shape. Each of the three germ layers (ectoderm, mesoderm amd endoderm) forms specific tissues and organs in the developing embryo. The endometrium and part of the blastocyst mesh and develop in the placenta which allows for the passages of nutrients, oxygen and waste to and from baby and mother.

Foetal development

Significant growth and development takes place within the first three months following conception.

Month 1

At the end of the first month the embryo has a definite form. The head, trunk and the tiny buds that will become the arms and legs are visible. The cardiovascular system has begun to function and the umbilical cord is visible in its most primitive form.

Month 2

During the second month the embryo grows to 2.5 cm in length. The head and facial features develop as the eyes, ears, nose, lips, tongue and tooth buds form. The arms and legs also take shape. Although the gender of the foetus is not yet visible, all external genitalia are present. Cardiovascular function is complete and the umbilical cord has a definite form. From the eighth week the embryo is called a foetus.

Month 3

During the third month the foetus grows to 7.5 cm in length. Teeth and bones begin to appear and the kidneys start to function. The foetus opens its mouth to swallow, grasps with its fully developed hands and prepares for breathing by inhaling and exhaling amniotic fluid (although its lungs are functioning properly). At the end of the third month, or trimester, the foetus' gender is distinguishable.

Months 4 to 9

Over the remaining six months, the foetal growth continues as internal and external structures develop at a rapid rate. In the third trimester the foetus stores the fats and minerals it will need to live outside the womb. At birth the average full-term foetus measures 51 cm and weighs 3 to 4 kgs.

Birth

Childbirth is divided into three stages. The duration of each stage varies according to the size of the uterus, the woman's age and the number of previous pregnancies. The first stage of labour is when the foetus begins its descent and the cervix begins to dilate to prepare to allow the foetus to pass from the uterus into the vagina. During this stage the amniotic sac ruptures as the uterine contractions increase in frequency and intensity (the amniotic sac can also rupture before the onset of labour).

The second stage of labour begins with full cervical dilation and ends with delivery of the foetus. The third stage of labour starts immediately after childbirth and ends with the placenta expulsion. After the neonate is delivered the uterus continues to contract intermittently and grows smaller.

Female reproductive changes with ageing

Declining oestrogen and progesterone levels cause numerous physical changes in women with age. Ovulation usually stops one to two years before the menopause. As the ovaries reach the end of their productive cycle, they become unresponsive to gonadotrophic stimulation. With ageing the ovaries atrophy and become thicker and smaller.

The vulva also atrophies with age and the tissue shrinks. Atrophy causes the vagina to shorten and the mucous lining to become thin, dry and less elastic. After the menopause the uterus shrinks rapidly to half its premenstrual weight. The cervix atrophies and no longer produces mucus for lubrication and the endometrium and myometrium become thinner. In the breasts the glandular, supporting and fatty tissues atrophy and as the Cooper's ligaments lose their elasticity the breasts become pendulous.

Anatomy of the female breast

The female breasts are accessory organs to the female reproductive system and their function is to produce and secrete milk after pregnancy.

Position of the breast

The breasts lie on the pectoral region of the front of the chest. They are situated between the sternum and the axilla, extending from approximately the second to the sixth rib. The breasts lie over the pectoralis major and serratus anterior muscles and are attached to them by a layer of connective tissue.

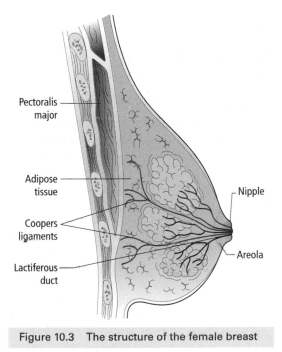

Pectoralis major

Adipose tissue

Coopers ligaments

Lactiferous duct

Nipple

Areola

Figure 10.3   The structure of the female breast

Structure of the breast

The breasts consist of glandular tissue arranged in lobules, supported by connective, fibrous and adipose tissue. The lobes are divided into lobules which open up into milk ducts.

The milk ducts open into the surface of the breast at a projection called the nipple. Around each nipple the skin is pigmented and forms the areola. This varies in colour from a deep pink to a light or dark brown colour. A considerable amount of fat or adipose tissue covers the surface of the gland and is found between the lobes. The skin on the breast is thinner and more translucent than body skin.

Support of the breast

The breasts are supported and slung in powerful suspensory Cooper's ligaments which go around the breast, with both ends being attached to the chest wall. The pectoralis major and serratus anterior muscles help to support the ligaments. If the breast grows large due to adolescence or pregnancy, the Cooper's ligaments may become irreparably stretched and the breast will then sag. With age, the supporting ligaments, along with the skin and the breast tissue, become thin and inelastic and the breasts lose their support.

Physiology of the breast

Lymphatic drainage

The breasts contain many lymphatic vessels and the lymph drainage is very extensive, draining mainly into the axillary nodes under the arms.

Blood supply

The blood vessels supplying blood to the breasts include the subclavian and axillary arteries.

Nerve supply

There are numerous sensory nerve endings in the breast, especially around the nipple. When these touch receptors are stimulated in lactation, the impulses pass to the hypo-thalamus and the flow of the hormone oxytocin is increased from the posterior lobe of the pituitary. This promotes the constant flow of milk when required.

Hormones

The hormones responsible for developing the breast are:

- **Oestrogen** – this is responsible for the growth and development of the secondary sex characteristics.

- **Progesterone** – this causes the mammary glands to increase in size if fertilisation and subsequent pregnancy occurs.

Development of the breasts

Puberty

The breast starts out as a nipple which projects from the surrounding ring of pigmented skin called the areola. Approximately two or three years before the onset of menstruation, the fat cells enlarge in response to the sex hormones (oestrogen and progesterone) released during adolescence.

Key note

The breasts change monthly in response to the menstrual cycle. The action of the female hormone progesterone increases blood flow to the breasts, which increases fluid retention and the breasts may increase in size, causing them to feel swollen and uncomfortable.

Pregnancy

During pregnancy the increased production of oestrogen and progesterone causes an increase in blood flow to the breasts. This causes an enlargement of the ducts and lobules of the breasts in preparation for lactation and there is an increase in fluid retention. The areolae and the nipples enlarge and become more pigmented.

Menopause

The reduction in the female hormones during the menopause causes the glandular tissue in the breast to shrink and the supporting ligaments, along with the skin, become thinner and lose their elasticity. Therefore, during the menopause the breasts begin to lose their support and uplift, although the degree of loss is dependent on the original strength of the suspensory ligaments.

Factors determining size and shape

The size of the breast is largely determined by genetic factors, although there are other factors such as:

- amount of adipose tissue present
- fluid retention
- level of ovarian hormones in the blood and the sensitivity of the breasts to these hormones
- degree of ligamentary suspension
- exercise undertaken.

Key note

Exercise may help to strengthen the pectoral muscles which will help to support the ligaments and increase their uplift. However, if the wrong type of exercise is undertaken and sufficient support is not provided for the breasts during exercise, the ligaments may become irreparably stretched.

Male reproductive system

The male reproductive system consists of the:

- **testes**
- **epididymides**
- **vas deferentia**
- **prostate gland**
- **ejaculatory ducts**
- **Cowper's glands**
- **urethra**
- **penis**.

Testes

The testes are the reproductive glands of the male and lie in the scrotal sac. Each testis consists of approximately 200 to 300 lobules. These are separated by connective tissue and are filled with seminiferous tubules in which sperm cells are formed. Between the tubules are a group of secretory cells known as the interstitial cells which produce male sex hormones.

The testes are specialised to produce and maintain sperm cells and produce male sex hormones known collectively as androgens. Testosterone is the most important androgen as it stimulates the development of the male reproductive organs. It is also responsible for the development and maintenance of the male secondary sexual characteristics.

Epididymides

The epididymides are coiled tubes leading from the seminiferous tubules of the testis to the vas deferens. They store and nourish immature sperm cells and promote their maturation until ejaculation.

Vas deferentia

The vas deferentia are tubes leading from the epididymis to the urethra and through which the sperm are released.

Key note

The vas deferentia are cut in the operation known as a vasectomy which produces sterilisation in the male.

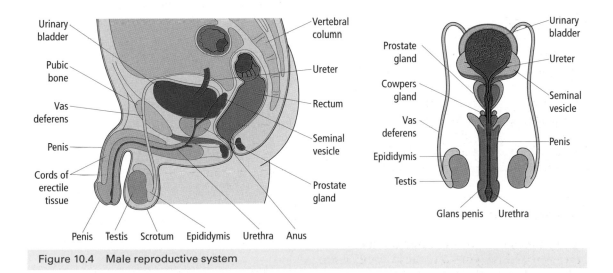

Figure 10.4   Male reproductive system

Seminal vesicles

The seminal vesicles are pouches lying on the posterior aspect of the bladder, attached to the vas deferens. They secrete an alkaline fluid which contains nutrients and is added to sperm cells during ejaculation.

Ejaculatory ducts

The two ejaculatory ducts are short tubes which join the seminal vesicles to the urethra.

Cowper's glands

The Cowper's glands are a pair of small glands that open into the urethra at the base of the penis. These glands produce further secretions to contribute to the seminal fluid, but less than that of the prostate gland or seminal vesicles.

Prostate gland

The prostate gland is a male accessory gland about the size of a walnut. It lies in the pelvic cavity in front of the rectum and behind the symphysis pubis. During ejaculation it secretes a thin, milky fluid that enhances the mobility of sperm and neutralises semen and vaginal secretions.

Key note

The prostate gland commonly becomes enlarged in older men, causing difficulty in passing urine due to constriction of the urethra.

Urethra

The urethra provides a common pathway for the flow of urine and the secretion of semen. A sphincter muscle prevents both functions occurring at the same time.

Penis

The penis is composed of erectile tissue and is richly supplied with blood vessels. When stimulated by sexual activity the blood vessels become engorged with blood and the penis becomes erect. Its function is to convey urine and semen.

Male reproductive changes with ageing

Physiological changes in older men include reduced testosterone production, which in turn may cause decreased libido. A reduced testosterone level also causes the testes to atrophy and soften and decreases sperm production by around 48 to 69 per cent between the ages of 60 and 80. Normally, the prostate gland enlarges with age and its secretions diminish. Seminal fluid also decreases in volume and becomes less viscous.

Disorders of the female reproductive system

Amenorrhea

Amenorrhea is the absence or stopping of the menstrual periods. Its causes may be associated with the deficiency of ovarian, pituitary or thyroid hormones, mental disturbances, depression, radical weight loss, stress, excessive exercise or a major change in surroundings or circumstances.

Cancer of the breast

Most breast cancers are detected by the woman noticing a breast or axillary lump. Mammography screening can confirm these lumps. Breast cancer can present as redness and pain, discharge from or retraction of the nipple. Cancer can spread locally or to the axilla and neck lymph nodes, causing oedema of the arm, or by blood to the lung, bone and liver. The type of breast cancer can determine whether the spread is rapid or very slow.

Cancer of the cervix

It is asymptomatic in the early stages. Later there may be a foul-smelling, bloodstained discharge through the vagina. Lower back pain, loss of weight, unexplained anaemia and pain during intercourse are other symptoms.

Cancer of the ovaries

It is asymptomatic. Diagnosis is usually made after the cancer has spread extensively. The symptoms are vague and are usually associated with gastrointestinal symptoms such as bloating of the abdomen, mild abdominal pain and excessive passage of gas. There may be fluid in the peritoneal cavity in the late stages. Hormone changes may result in abnormal vaginal bleeding.

Dysmenorrhea

This condition is defined as painful and difficult menstruation. It presents with spasms and congestion of the uterus, resulting in cramping, lower abdominal pains which start before or with the menstrual flow and continue during menstruation. It is often associated with nausea, vomiting, headache and a feeling of faintness.

Ectopic pregnancy

This is the development of a foetus at a site other than in the uterus. An ectopic pregnancy may occur if the fertilised egg remains in the ovary or fallopian tube, or if it lodges in the abdominal cavity. The most common type of ectopic pregnancy occurs in the fallopian tube. There is a danger of haemorrhage and death as growth of the foetus may cause the tube to rupture and bleed.

Endometriosis

Condition in which tissue resembling the lining of the uterus (endometrium) is abnormally present in the pelvic cavity.

Fibroid

This condition is an abnormal growth of fibrous and muscular tissue, one or more of which may develop in the muscular wall of the uterus. Fibroids can cause pain and excessive bleeding and become extremely large. Although they do not threaten life they render pregnancy unlikely. Some fibroids may be removed surgically but in other cases a hysterectomy may be necessary.

Infertility

This is the inability in a woman to conceive or in a man to induce conception. Female infertility may be due to a failure to ovulate, obstruction of the fallopian tubes or endometriosis.

Polycystic ovary syndrome (also known as Stein-Leventhal syndrome)

This is a hormonal disorder in which there is inadequate secretion of the female sex hormones. As a result, the ovarian follicles fail to ovulate and remain as multiple cysts distending the ovary. Other associated symptoms include obesity, hirsutism, acne and infertility.

Premenstrual syndrome

Premenstrual syndrome is a term for the physical and psychological symptoms experienced 3 to 14 days prior to the onset of menstruation. The condition presents with varying symptoms such as headache, bloatedness, water retention, backache, changes in coordination, abdominal pain, swollen and painful breasts, depression, irritability and craving for sweet things.

Disorders of the male reproductive system

Cancer of the testis

The first symptom is a slight enlargement of the testis. It may be accompanied by pain, discomfort and heaviness of the scrotum. Soon there is a rapid enlargement of the testis which can become hot and red.

Cancer of the prostate

Usually no symptoms are seen. If the cancer is located close to the urethra there may be a frequency of micturition, an urgency or difficulty in voiding, blood in the urine or blood when ejaculating. Cancer of the prostate is often diagnosed by rectal examination where it feels nodular and hard. Prostate cancer usually spreads to the bones and produces bony pain or causes fractures in the bone after a trivial injury. In the advanced stage, as in all cancers, the person loses weight and is anaemic.

Infertility

Causes of male infertility can include decreased numbers or mobility of sperm or may be due to the total absence of sperm. In both male and female infertility, the cause may also be associated with stress.

Prostatitis

Prostatitis is the inflammation of the prostate gland which is usually caused by bacteria. This condition presents with a frequency and urgency on passing urine (urine may be cloudy). High fever with chills, muscle and joint pain are common. A dull ache may be present in the lower back and pelvic area.

Interrelationships with other systems

The reproductive system links to the following body systems:

Cells and tissues

Ova are the reproductive cells in the female and sperm cells are the reproductive cells in the male.

Skeletal

The pelvis offers protection for the uterus.

Muscular

Smooth muscle is responsible for the passage of ova from the ovaries to the vagina and sperm from the testes to the urethra. During orgasm in the female the muscles of the perineum, uterine wall and the uterine tubes contract rhythmically. During orgasm in the male motor impulses are transmitted to skeletal muscles at the base of the erectile penis, causing them to contract rhythmically.

Circulatory

During erection of the penis the vascular spaces within the erectile tissue become engorged with blood as arteries dilate and veins are compressed. During periods of sexual stimulation the erectile tissues of the clitoris become engorged with blood.

Nervous

Orgasm is the culmination of sexual stimulation. The movement of semen occurs as a result of sympathetic reflexes.

Endocrine

The ovaries in women and the testes in the male are responsible for the development of the secondary sexual characteristics.

Key words associated with the reproductive system

ovaries	progesterone	Cowper's gland
fallopian tube	pregnancy	Cooper's ligament
uterus	menopause	penis
vagina	testes	scrotum
vulva	epididymides	sperm
genitalia	vas deferens	testosterone
breast/mammary gland	ejaculatory ducts	seminiferous tubules
ovum	urethra	ejaculation
ova	seminal vesicles	semen
oestrogen	prostate	

Summary of the reproductive system

- The male and female reproductive systems function to produce the sex hormones responsible for the male and female characteristics and the cells required for reproduction.

- The structures of the female reproductive system include the **ovaries, fallopian tubes, uterus, vagina** and **vulva.**

- The **breasts,** or mammary glands, are considered to be part of the female reproductive system.

- The **ovaries** lie on the lateral walls of the pelvis and have two distinct functions – the production of **ova** and the secretion of the female hormones **oestrogen** and **progesterone**.

- The **fallopian tubes** transport **ova** from the **ovaries** to the **uterus**.

- The **uterus** is situated behind the bladder and in front of the rectum and is designed to receive, nourish and protect a fertilised **ovum**.

- The **vagina** is a muscular and elastic tube, designed for the reception of sperm and to provide a passageway for menstruation and childbirth.

- The **vulva** is a collective term for the female **genitalia**.

- In the female the levels of **oestrogen** and **progesterone** decrease with age.

- Ovaries atrophy and become thicker and smaller with age.

- The **vulva** atrophies and tissue shrinks with age. Atrophy causes the **vagina** to shorten and the mucous lining to become thin, dry and less elastic.

- After the **menopause** the **uterus** shrinks rapidly to half its premenstrual weight.

- The **breasts** atrophy and lose their elasticity and support.

- **Pregnancy** starts with fertilisation and ends with childbirth, and consists of the following stages – fertilisation, pre-embryonic development, formation of embryo, foetal development and birth.

- The structures of the male reproductive system include the **testes, epididymides, vas deferens, ejaculatory ducts, urethra, seminal vesicles, prostate, Cowper's glands** and **penis.**

- The **testes** lie in a scrotal sac. They produce and maintain **sperm** cells and produce the male sex hormone **testosterone.**

- Each **testis** is filled with **seminiferous tubules** in which **sperm** cells are formed.

- The **epididymides** are coiled tubes that lead from the **seminiferous tubules** of the **testis** to the **vas deferens.** They store and nourish immature **sperm** cells and promote their maturation until **ejaculation.**

- The **vas deferentia** lead from the **epididymides** to the **urethra** and are tubes through which the **sperm** are released.

- The **seminal vesicles** are pouches lying on the posterior aspect of the bladder, attached to the vas deferens. They secrete an alkaline fluid which contains nutrients and is added to **sperm** cells during **ejaculation.**

- The two **ejaculatory ducts** are short tubes which join the **seminal vesicles** to the **urethra.**

- The **Cowper's glands** are a pair of small glands that open into the **urethra** at the base of the **penis.** These glands produce further secretions to contribute to the **seminal fluid.**

- The **prostate** gland lies in the pelvic cavity in front of the rectum and behind the symphysis pubis. During ejaculation it secretes a thin, milky fluid that enhances the mobility of sperm and neutralises **semen** and vaginal secretions.

- The **urethra** provides a common pathway for the flow of urine and the secretion of **semen.**

- The **penis** is composed of erectile tissue and is richly supplied with blood vessels. Its function is to convey urine and **semen.**

- In the male decreased levels of **testosterone** decreases sexual desire and viable **sperm.** The testes also atrophy as muscle strength decreases.

Multiple-choice questions

1. The main function of the ovaries is to:

a   accommodate a growing foetus during pregnancy

b   serve as a site for fertilisation

c   produce mature ova

d   receive male sperm

2. Once an ovum has become fertilised it is known as a:

a   blastocyst

b   zygote

c   embryo

d   foetus

3. A foetus' gender is distinguishable at:

a   eight weeks

b   the end of the first month

c   the end of the first trimester

d   a month before birth

4. The function of the fallopian tubes is to:

a   convey the ovum from the ovary to the uterus

b   convey the ovum from the ovary to the vulva

c   prepare for the implantation of a fertilised ovum

d   secrete mucus

5. The uterus is situated:

a   in front of the bladder and behind the rectum

b   behind the bladder and in front of the rectum

c   on the lateral walls of the pelvis

d   at the entrance of the vulva

6. The cervix is:

a   a thick, muscular structure that opens into the vagina

b   an outer covering of the uterus

c   the largest and main part of the uterus

d   the dome-shaped part of the uterus

7. The inner mucous membrane lining of the uterus is called the:

a   perimetrium

b   perineum

c   myometrium

d   endometrium

8. Which of the following statements is true?

a   the seminal vesicles secrete an alkaline fluid which contains bacteria

b   vas deferentia are the tubes through which sperm is released

c   the prostate gland lies in front of the symphysis pubis and behind the rectum

d   a male urethra can only serve as a pathway for semen

9. Where in the male reproductive system are sperm cells stored to maturation?

a   vas deferentia

b   penis

c   epididymides

d   Cowper's glands

10. The collective term for male hormones is:

a   gonads

b   androgens

c   vesicles

d   none of the above

11. Which of the following hormones prepares the lining of the uterus for the implantation of a fertilised egg?

a   progesterone

b   oestrogen

c   follicle-stimulating hormone

d   luteinising hormone

12. Where does fertilisation of the ovum take place?

a   ovaries
b   uterus
c   fallopian tubes
d   vagina

13. The function of the vagina is to:

a   receive sperm
b   provide a passageway for childbirth and menstruation
c   provide an acid environment to prevent the growth of microbes
d   all of the above

14. The absence or stopping of menstrual periods is known as:

a   endometriosis
b   dysmenorrhea
c   amenorrhea
d   premenstrual syndrome

the
digestive
system

Introduction

In the digestive system food is broken down and made soluble before it can be absorbed by the body for nutrition. Food is taken in through the mouth, broken into smaller particles and absorbed into the bloodstream where it is utilised by the body. Waste materials not required by the body are then passed through the body to be eliminated. Once food has been absorbed by the body, it is converted into energy to fuel the body's activities. This is known as metabolism.

Objectives

By the end of this chapter you will be able to recall and understand the following knowledge:

■ the functions of the digestive system

■ the process of digestion from the ingestion of food to the elimination of waste

■ the structure and functions of the organs associated with digestion

■ the absorption of nutrients and their utilisation in the body

■ the sources and functions of the main food groups required for good health

■ the interrelationships between the digestive and other body systems

■ disorders of the digestive system.

Functions of the digestive system

The digestive system serves two major functions:

- the breaking down of food and fluid into simple chemicals that can be absorbed into the bloodstream and transported throughout the body
- the elimination of waste products through excretion of faeces via the anal canal.

The structure and function of digestive organs

Digestion occurs in the alimentary tract, which is a long, continuous, muscular tube extending from the mouth to the anus. The process of breaking down food is called digestion. Digestion involves the following processes:

Ingestion

This is the act of taking food into the alimentary canal through the mouth.

Digestion

- **Mechanical digestion** – this is the breaking down of solid food into smaller pieces by the chewing action of the teeth, known as mastication, and the churning action of the stomach, assisted by peristalsis.

- **Chemical digestion** – this involves the breakdown of large molecules of carbo-hydrates, proteins and fats into smaller ones by the action of digestive enzymes.

Absorption

This is the movement of soluble materials out through the walls of the small intestine. Nutrients are absorbed through the villi and pass out into the network of blood and lymph vessels to be delivered to various parts of the body.

Assimilation

This is the process by which digested food is used by the tissues after absorption.

Elimination/defecation

This is the expulsion of the semi-solid waste called faeces through the anal canal.

The digestive system consists of the following parts:

- **mouth**
- **pharynx**
- **oesophagus**
- **stomach**
- **small intestine** (consisting of the duodenum, jejenum and ileum)
- **large intestine** (consisting of the caecum, appendix, colon and rectum)
- **anus**.

The **pancreas**, **gall bladder** and the **liver** are accessory organs to digestion.

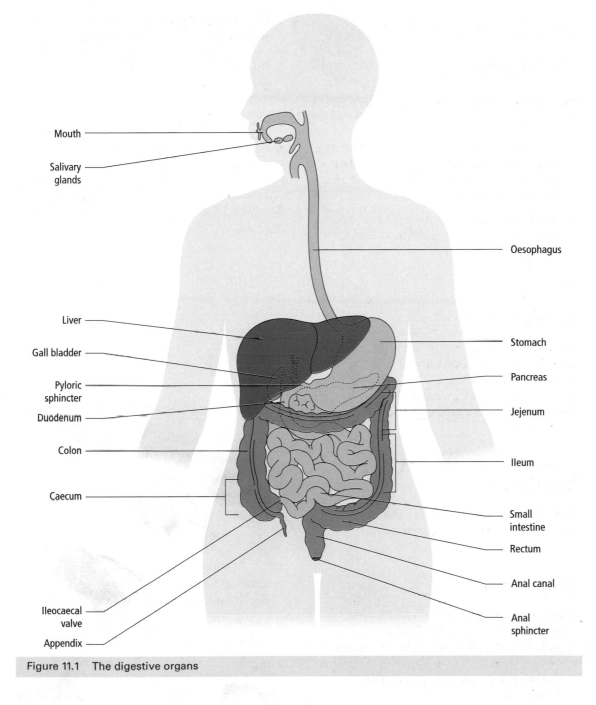

Figure 11.1   The digestive organs

Mouth

The digestive system commences in the mouth. Food is broken up into smaller pieces by the action of the jaws and the teeth, and shaped into a ball by the tongue. Mastication renders the food small enough to be swallowed and allows saliva to be thoroughly mixed with it.

The smell and sight of food triggers the reflex action of the secretion of saliva in the mouth. Saliva enters the mouth from three pairs of salivary glands. These are the:

■ **sublingual glands** – located in the lower part of the mouth on either side of the tongue

■ **submandibular glands** – located inside the arch of the mandible

■ **parotid glands** – located superficial to the masseter muscle.

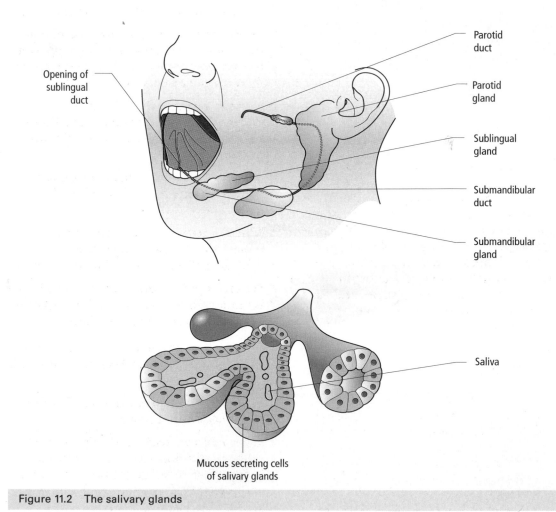

Figure 11.2   The salivary glands

Saliva, containing the enzyme salivary amylase, commences the digestion of starch, or carbohydrates, in the mouth.

Key note

An enzyme is a chemical catalyst which activates and speeds up a chemical reaction without any change to itself. Enzymes are highly specific in that each enzyme catalyses only one type of metabolic action. An example is salivary amylase, which will only act on starch and has no effect on protein.

Pharynx and oesophagus

The ball of food is projected to the back of the mouth. The muscles of the pharynx force the food down the oesophagus, which is a long narrow tube linking the pharynx to the stomach. A lubricative substance called mucus, secreted from the lining of the oesophagus, makes the food easier to swallow. The food is then conveyed by peristalsis down the oesophagus to the stomach.

Key note

Peristalsis is the coordinated, rhythmical contraction of the circular and oblique muscles in the wall of the alimentary tract. These muscles work in opposition to one another to break down food and move it along the alimentary canal. Peristalsis is an automatic action stimulated by the presence of food and occurs in all sections of the alimentary tract.

Stomach

The stomach is a curved, J-shaped, muscular organ, positioned in the left-hand side of the abdominal cavity, below the diaphragm.

Food enters the stomach via the cardiac sphincter, which is a strong, circular muscle at the junction of the stomach and the oesophagus. Its function is to control the entry of food into the stomach.

The stomach consists of five layers:

- **Peritoneum** – this is a serous membrane that lines the abdominal cavity, supporting the alimentary canal, and secretes a serous fluid which prevents friction.

- **Muscular coat** – this consists of longitudinal, circular and oblique fibres which assist the mechanical breakdown of food.

- **Sub-mucous coat** – this is an areolar tissue containing blood vessels and lymphatics.

- **Mucous coat** – this secretes mucus to protect the stomach lining from the damaging effects of the acidic gastric juice.

- **Surface epithelium** – in the stomach this is infolded into numerous tubular gastric glands which secrete gastric juice.

The main constituents of gastric juice, produced and secreted by cells in the stomach wall, are:

- **Pepsin** – this is an enzyme which starts the breakdown of proteins.

- **Hydrochloric acid** – this provides the acidic conditions needed for pepsin to become active, kill germs present in food and prepare it for intestinal digestion.

- **Mucus** – this is secreted by the neck cells in the stomach wall. It protects the stomach lining from the damaging effects of the acidic gastric juice.

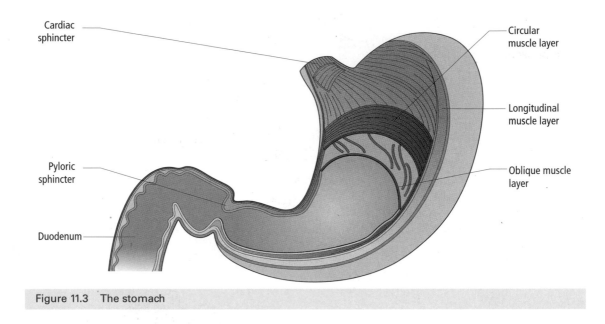

Cardiac sphincter

Circular muscle layer

Longitudinal muscle layer

Oblique muscle layer

Pyloric sphincter

Duodenum

Figure 11.3   The stomach

- **Rennin** – this is an enzyme found in the gastric juices of infants that curdles milk protein.

- **Gastrin** – this is a hormone released by endocrine cells in the stomach wall and is stimulated by the presence of food. This hormone circulates in the bloodstream, stimulating the further release of gastric juice.

The functions of the stomach are to:

- churn and break up large particles of food mechanically

- mix food with gastric juice to begin the chemical breakdown of food

- commence the digestion of protein

- absorb alcohol.

Food stays in the stomach for approximately five hours until it has been churned down to a liquid state called chyme. Chyme is then released at intervals into the first part of the small intestine. The exit from the stomach is controlled by the pyloric sphincter, which sits at the junction of the stomach and the duodenum. Its function is to relax and release chyme at intervals into the small intestine.

Small intestine

The small intestine is about 7 m long and consists of three parts:

- **duodenum** – the first part of the small intestine

- **jejenum**

- **ileum** – where the main absorption of food takes place.

The small intestine consists of four layers the same as the stomach:

- peritoneum

- muscular coat excluding the oblique fibres

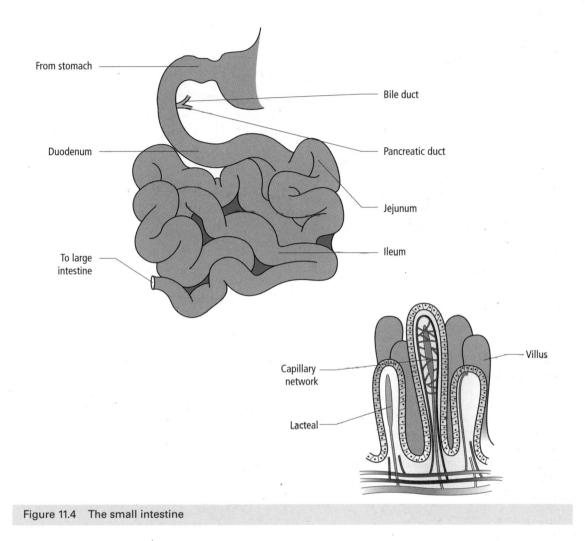

From stomach

Bile duct

Duodenum

Pancreatic duct

Jejunum

Ileum

To large intestine

Capillary network

Villus

Lacteal

Figure 11.4   The small intestine

- sub-mucous layer containing numerous blood and lymph vessels and nerves
- circular folds of mucosa which protect the intestine from bacteria.

The special features of the small intestine are the thousands of minute projections called villi, each containing a lymph vessel called a lacteal. The villi have a network of capillaries into which the nutrients pass to be absorbed into the bloodstream.

Chemical breakdown of food

The muscles in the wall of the small intestine continue the mechanical breakdown of food by peristaltic movements, while the chemical digestion is brought about by the following juices which prepare the food to be absorbed into the bloodstream.

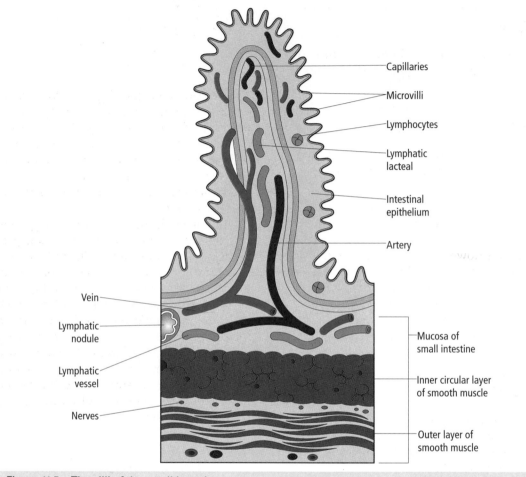

Capillaries

Microvilli

Lymphocytes

Lymphatic
lacteal

Intestinal
epithelium

Artery

Vein

Lymphatic
nodule

Lymphatic
vessel

Nerves

Mucosa of
small intestine

Inner circular layer
of smooth muscle

Outer layer of
smooth muscle

Figure 11.5   The villi of the small intestine

Bile

Bile is a green, alkaline liquid produced in the liver and stored in the gall bladder. It consists of water, mucus, bile pigments, bile salts and cholesterol and is released at intervals from the bile duct when food enters the duodenum. The function of bile is to neutralise the chyme and break up any fat droplets in a process called emulsification.

Pancreatic juice

This is produced by the pancreas, a gland extending from the loop of the duodenum to behind the stomach. The pancreas secretes pancreatic juice into the duodenum and the enzymes contained within it continue the digestion of protein, carbohydrates and fats.

Trypsin

This continues the breakdown of proteins which was started by pepsin in the stomach.

Pancreatic amylase

This continues the breakdown of starch and has the same effects as salivary amylase.

Pancreatic lipase

This breaks down lipids into fatty acids and glycerol. The pancreas also has an endocrine function in that it secretes insulin from the islet of Langerhans cells in the pancreas. Insulin is important to digestion because it regulates blood sugar levels.

Intestinal juice

This is released by the glands of the small intestine and completes the final breakdown of nutrients, including simple sugars to glucose and protein to amino acids.

Carbohydrate digestion is completed by the following enzymes:

- **Maltase** – this splits maltose into glucose.

- **Sucrase** – this splits sucrose into glucose and fructose.

- **Lactase** – this splits lactose into glucose and galactose.

Protein digestion is completed by peptidases which split short-chain **polypeptides** into **amino acids.**

Absorption of the digested food

The absorption of the digested food takes place in the jejenum and mainly in the ileum. It occurs by diffusion through the villi of the small intestine, which are well supplied with blood capillaries to allow the digested food to enter. Each villus contains a lymph vessel called a **lacteal** into which **fatty acids** and **glycerol** can pass.

Simple sugars from carbohydrate digestion and amino acids from protein digestion pass into the bloodstream via the villi and are then carried to the liver, via the hepatic portal vein, to be processed. The products of fat digestion pass into the intestinal lymphatics, which absorb the fat molecules and carry them through the lymphatic system before they reach the blood circulation. Vitamins and minerals travel across to the blood capillaries of the villi and are absorbed into the bloodstream to assist in normal body functioning and cell metabolism.

How the body's nutrients are assimilated

Once all the nutrients have been absorbed into the bloodstream they are transported to the body's cells for metabolism:

- **Glucose** – this is the end product of carbohydrate digestion and is used to provide energy for the cells to function.

- **Amino acids** – these are the end products of protein digestion. They are used to produce new tissues, repair damaged cell parts and to formulate enzymes, plasma proteins and hormones.

- **Fatty acids and glycerol** – these are the end products of fat digestion. Fats are used primarily to provide heat and energy in addition to glucose. Those fats which are not required immediately by the body are used to build cell membranes and some are stored under the skin or around vital organs such as the kidneys and the heart.

When all the body's nutrients have been assimilated by the body, the fate of the undigested food is to pass into the large intestine where it is eventually eliminated from the body.

Large intestine

The large intestine is formed of the **caecum**, **appendix**, **colon** and **rectum**. It coils around the small intestine and is characterised by:

- three bands of longitudinal muscle

- deep, longitudinal folds of mucosa which increase in the rectum

- numerous tubular glands which secrete mucus from their goblet cells.

The **caecum** is a small pouch to which the appendix is attached and into which the ileum opens through the ileocaecal valve.

The **appendix** has no known function in humans.

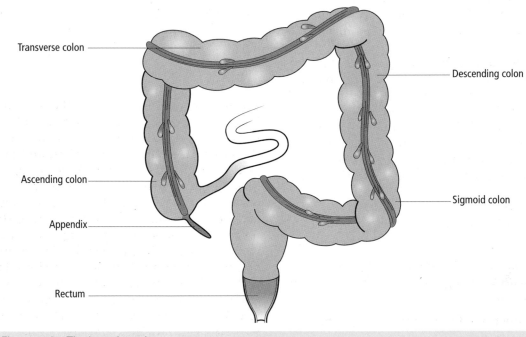

Transverse colon

Descending colon

Ascending colon

Sigmoid colon

Appendix

Rectum

Figure 11.6   The large intestine

The **colon** is the main part of the large intestine and is divided into four sections – ascending, transverse, descending and sigmoid colons:

- **Descending colon** – this is the part that passes upwards on the right side of the abdomen from the caecum (a pouch at the junction of the small and large intestines) to the lower edge of the liver.

- **Transverse colon** – this is the longest and most mobile part and extends across the abdomen from right to left, below the stomach.

- **Descending colon** – this is the part that passes downwards along the left side of the abdominal cavity to the brim of the pelvis.

- **Sigmoid colon** – this is the S-shaped part of the large intestine between the descending colon and the rectum.

The **rectum** is the last part of the large intestine. It is about 12 cm long and runs from the sigmoid colon to the anal canal. It is firmly attached to the sacrum and ends about 5 cm below the tip of the coccyx, where it becomes the anal canal. Faeces are stored in the rectum before defecation.

The functions of the large intestine are:

- absorption of most of the water from the faeces in order to conserve moisture in the body

- formation and storage of faeces which consist of undigested food, dead cells and bacteria

- production of mucus to lubricate the passage of faeces

- the expulsion of faeces out of the body through the anus.

Anus

The anus is an opening at the lower end of the alimentary canal. It is the anal canal through which faeces are discharged. The anus is guarded by two sphincter muscles:

- **Internal sphincter** – this is composed of smooth muscle under involuntary control

- **External sphincter** – this is composed of skeletal muscle under voluntary control.

The anus remains closed except during defecation.

Accessory organs to digestion

Liver

The liver is the largest gland in the body and is situated in the upper right-hand side of the abdominal cavity, under the diaphragm. It has a soft, reddish-brown colour and four lobes. Its internal structure is made up of cells called hepatocytes. The liver receives oxygenated blood from the hepatic artery and deoxygenated blood from the hepatic portal vein. Blood from the digestive tract, which is carried in the portal veins, brings newly absorbed nutrients into the sinusoids and nourishes the liver cells. The liver is a vital organ and, therefore, has many important functions in the metabolism of food. It regulates the nutrients absorbed from the small intestine to make them suitable for use in the body's tissues.

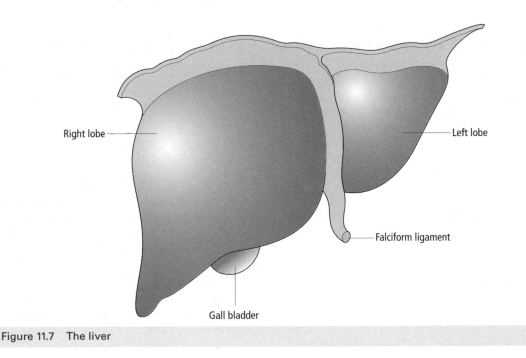

Figure 11.7   The liver

Functions of the liver

Secretion of bile

Bile is manufactured by the liver, but is stored and released by the gall bladder to assist the body in the breakdown of fats.

Regulation of blood sugar levels

When the blood sugar levels rise after a meal the liver cells store excess glucose as glycogen. Some glucose may be stored in the muscle cells as muscle glycogen. When both these stores are full surplus glucose is converted into fat by the liver cells.

Regulation of amino acid levels

As our bodies cannot store excess protein and amino acids they are processed by the liver. Some are removed by the liver cells and used to make plasma proteins. Some are left for the body cell tissues' use, while the rest are deaminated and excreted as urea in the kidneys.

Regulation of the fat content of blood

The liver is involved in the processing and transporting of fats. Those already absorbed in the diet are used for energy and excess fats are stored in the tissues.

Regulation of plasma proteins

The liver is active in the breakdown of worn-out red blood cells.

Detoxification

The liver detoxifies harmful toxic waste and drugs and excretes them in bile or through the kidneys.

Storage

The liver stores vitamins A, D, E, K and B12 and the minerals iron, potassium and copper. The liver can also hold up to a litre of blood. During exercise the liver supplies extra blood and increases oxygen transport to the muscles.

The production of heat

Due to its many functions, the liver generates heat.

Gall bladder

The gall bladder is a pear-shaped organ attached to the posterior and inferior surface of the liver by the cystic and bile ducts.

Bile is a thick, alkaline liquid that is produced in the liver as a result of the breakdown of red blood cells. Bile is partially an excretory product and partially a digestive secretion. Bile salts (sodium and potassium) play a role in emulsification and the breakdown of large fat globules.

When worn-out red blood cells are broken down, substances such as iron and globin are recycled, but some of the bilirubin is excreted

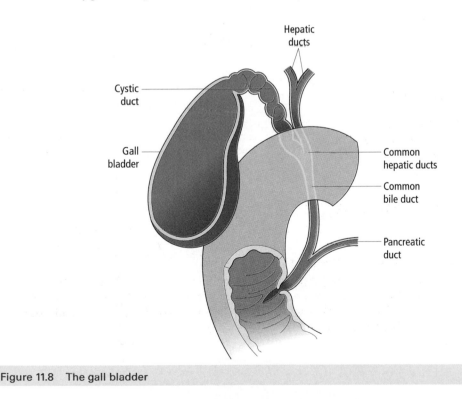

Figure 11.8　The gall bladder

into the bile ducts. Bilirubin is eventually broken down in the intestines and one of its breakdown products gives faeces the normal brown colour.

Functions of the gall bladder

The gall bladder stores and concentrates bile produced by the liver until it is needed. It releases bile into the common bile duct for delivery to the duodenum.

Pancreas

The pancreas is situated behind the stomach, between the duodenum and the spleen. It is divided into a head, body and tail. The head is the expanded portion that fits into the C-shaped curve of the duodenum. The pancreas is composed of numerous lobules, each containing secretory alveoli (small, saclike cavities) which contain cells that produce pancreatic juice. In between the network of alveoli are the islets of Langerhans which produce insulin.

Functions of the pancreas

The pancreas has two functions – exocrine and endocrine.

Function of exocrine

The pancreas secretes pancreatic juice which contains water, alkaline salts, the enzymes lipase, pancreatic amylase, trypsinogen and

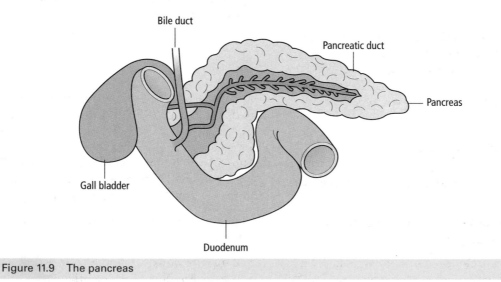

Figure 11.9 The pancreas

chymotrypsinogen. The alkalinity of pancreatic juice helps to neutralise the acidity of chyme from the stomach and allows the pancreatic and intestinal enzymes to work.

Function of endocrine

The islets of Langerhans are endocrine glands which secrete the hormone **insulin** into the bloodstream. The insulin circulates around the body in the blood and is important to carbohydrate metabolism.

Key note

Without insulin no glycogen can be stored in the liver and muscles and glucose cannot be oxidised to produce energy.

Nutrition

Nutrition is the utilisation of food to facilitate growth and maintain the normal working of the body. Poor nutrition can have a dramatic effect on our general health, energy levels, sleep patterns and stress response.

Carbohydrates

This group is also known as starches and sugars.

- Dietary sources – bread, cereals, potatoes, fruit and sugars.
- Main functions – carbohydrates are the body's main source of energy, required for the metabolism of other nutrients such as proteins and fats.

Proteins

- Dietary sources – first-class proteins such as fish, milk, eggs and meat. Second-class proteins include pulses, beans and peas.
- Main functions – protein is necessary for the growth and repair of the body tissues which are used in the production of hormones and enzymes.

Fats

Fats are classified as saturated or unsaturated, depending on whether they are solid (saturated) or liquid (unsaturated) at room temperature.

- Dietary sources – meat, milk, cheese, butter and eggs.
- Main functions – fats are a source of stored energy. They offer support and protection for the body and are used to build cell structures.

Water

Although water is not usually considered as food, it is nevertheless an essential nutrient needed by every part of the body. As our bodies are made up of more than 70 per cent water it is recommended that at least six to eight glasses of fresh water are consumed every day for efficient body functioning.

- Dietary sources – fresh water, fruit and vegetables.
- Main functions – water aids digestion and elimination. It is essential to maintaining the body's fluid balance and aids in the transport of substances around the body.

Fibre

Although fibre is not broken down into nutrients, it is a very necessary component for effective digestion.

- Dietary sources – sources in our diet include pulses, peas, beans, brown rice, wholemeal bread, jacket potatoes and green, leafy vegetables.

- Main functions – fibre aids digestion and bowel functioning. It provides the bulk in food to satisfy the appetite.

Vitamins

Vitamins are essential for normal physiological and metabolic functioning of the body. They regulate the body's processes and contribute to its resistance to disease. Vitamins are divided into two groups according to whether they are soluble in water or fat.

Vitamin A (fat-soluble)

- Dietary sources – carotene in carrots, liver, kidney, eggs, dairy products, fish and liver oils.

- Main functions – it is essential for healthy vision, healthy skin and mucous membrane.

Vitamin D (fat-soluble)

- Dietary sources – fish liver oils, fatty fish, margarine and eggs. It is also synthesised from ultraviolet light.

- Main functions – it is essential for healthy teeth and bones. It maintains the blood calcium level by increasing calcium absorption from food.

Vitamin E (fat-soluble)

- Dietary sources – peanuts, wheatgerm, milk, butter and eggs.

- Main functions – it inhibits the oxidation of fatty acids that help to form cell membranes.

Vitamin K (fat-soluble)

- Dietary sources – green, leafy vegetables, cereals, liver and fruit.

- Main functions – it is essential for blood clotting.

Vitamin B1 (water-soluble)

- Dietary sources – egg yolk, liver, milk, wholegrain cereals, vegetables and fruit.

- Main functions – it is necessary for the steady release of energy from glucose.

Vitamin B2 (water-soluble)

- Dietary sources – milk, liver, eggs and yeast.

- Main functions – it is essential for using energy released from food.

Vitamin B5 (water-soluble)

- Dietary sources – wholegrain cereals, yeast extract, liver, beans, nuts and meat.

- Main functions – it is involved in the breakdown of glucose to release energy.

Vitamin B6 (water-soluble)

- Dietary sources – wholegrain cereals, yeast extract, liver, meat, nuts, bananas, salmon and tomatoes.

- Main functions – it is necessary for the metabolism of protein and fat.

Vitamin B12 (water-soluble)

- Dietary sources – liver, kidney, milk, eggs and cheese.

- Main functions – it is necessary for the formation of red blood cells in bone marrow. It is also involved in protein metabolism.

Folic acid (water-soluble)

- Dietary sources – liver, kidney, leafy vegetables, oranges and bananas.

- Main functions – it is essential for the normal production of red and white blood cells.

Vitamin C (water-soluble)

- Dietary sources – citrus fruits and blackcurrants.

- Main functions – it assists in the formation of connective tissue and collagen. It helps prevent bleeding and aids healing.

Minerals

Minerals provide the body with materials for growth and repair and for the regulation of body processes. They are needed in trace amounts and are used to build bone, work muscles, support various organs and transport oxygen and carbon dioxide.

Calcium

- Dietary sources – milk, egg yolk, cheese and green, leafy vegetables.

- Main functions – it is essential for the formation of healthy bones and teeth, blood coagulation and the normal function of muscles and nerves.

Iron

- Dietary sources – liver, kidney, red meats, egg yolk, nuts and green vegetables.

- Main functions – it is essential for the production of haemoglobin in red blood cells.

Phosphorus

- Dietary sources – cheese, eggs, white fish, wholemeal bread, peanuts and yeast extract.

- Main functions – it is important in the formation of bones and teeth, muscle contraction and the transmission of nerve impulses.

Sulphur

- Dietary sources – egg yolk, fish, red meat and liver.

- Main functions – it is the main component of structural proteins (those in the skin and hair).

Sodium and chlorine

- Dietary sources – table salt, bacon, kippers and found in all body fluids.

- Main functions – they maintain fluid balance in the body. They are necessary for the transmission of nerve impulses and contraction of muscle.

Magnesium

- Dietary sources – green vegetables and salad.

- Main functions – it is important for the formation of bone and is required for the normal functioning of muscles and nerves.

Disorders of the digestive system

Anorexia nervosa

This is a psychological illness in which clients starve themselves or use other techniques, such as vomiting or laxatives, to induce weight loss. They are motivated by a false perception of their body image and a phobia of becoming fat. The result is a severe loss of weight with amenorrhea and even death from starvation.

Appendicitis

This is an acute inflammation of the appendix. The main symptom is abdominal pain centrally and in the right, lower abdomen, over the appendix. It is usually treated by surgical removal known as an appendectomy.

Bulimia

This is a psychological illness which is characterised by overeating (bingeing), followed by self-induced vomiting.

Cancer of the colon

In the early stages the signs and symptoms are vague and related to the location of the cancer. A dull abdominal pain may or may not be present. General symptoms include loss of weight, fatigue, anaemia and weakness. If the tumour is on the right side of the abdomen (caecum or ascending colon) symptoms of obstruction appear slowly, as tumours in this region generally tend to spread along the walls of the gut without narrowing the lumen of the gut. If on the left side (descending colon, sigmoid colon or rectum) the signs of obstruction appear early in the disease. There is constipation or diarrhoea with passage of pencil-shaped or ribbon-like stools. The blood in the stools may be red or dark in colour.

Cancer of the gall bladder

Indigestion and colicky pain may be present, especially after a fatty meal. The pain is located in the upper right quadrant of the abdomen and may be referred to the back, right shoulder, right scapula or between the scapulae.

Cancer of the liver

The more common type of cancer is that which has spread from other areas of the body – a metastatic carcinoma. Spread is common from those areas from which blood flows through the liver. Cancer can also arise from the liver tissue – primary cancer. Commonly, liver cancer is due to secondary spread from the stomach, intestine or pancreas. Liver cancer may be present as a swelling in the upper right quadrant associated with jaundice or fluid in the abdomen. Other general symptoms may include weight loss, weakness and loss of appetite. Usually, this type of cancer is well advanced when diagnosed, whether arising from the liver or secondary to cancer elsewhere in the body.

Cancer – oral

This may be caused by chronic irritation of the mucosa of the oral cavity, as in tobacco chewing. A recurrence of chronic ulcers of the mouth can lead to this type of cancer. Oral cancer may appear as a non-healing, slow-growing, red ulcer or as a growth. Usually it is painful and firm to touch.

Cancer of the pancreas

The person presents with severe weight loss and pain in the lower back. The pain increases a few hours after taking food and is worsened on lying down. If the tumour is growing around the bile duct obstruction may result in jaundice and diarrhoea. The accumulation of bilirubin under the skin causes severe itching. The jaundice may be so severe that the skin may turn green or black as the bilirubin changes in structure. The reduction in bile slows down the absorption and digestion of fat, causing clay-coloured, foul-smelling stools and diarrhoea. The cancer spreads directly and rapidly to the surrounding tissues, including the lymph nodes and liver. The kidneys, spleen and blood vessels may also be involved. The symptoms may vary according to the tissues affected.

Cancer of the stomach

In the early stages the person has chronic pain or discomfort in the upper part of the abdomen. Since the symptoms are vague, this cancer is often not diagnosed until it has spread considerably. There is weight loss, anaemia, loss of appetite and the person will feel easily fatigued. Vomiting is common and often contains blood. A mass may be felt in the upper abdomen. Indigestion and acidity is not relieved by medication.

Cirrhosis of the liver

Cirrhosis refers to a distorted or scarred liver as a result of chronic inflammation. The functional liver cells are replaced by fibrous or adipose connective tissue. The symptoms of cirrhosis include jaundice, oedema in the legs, uncontrolled bleeding and sensitivity to drugs. Cirrhosis may be caused by hepatitis, alcoholism, certain chemicals that destroy the liver cells or parasites that infect the liver.

Colitis

This is an inflammation of the colon. The usual symptoms are diarrhoea, sometimes with blood and mucus and lower abdominal pain.

Constipation

This condition presents as a difficulty in passing stools or where there is infrequent evacuation of the bowels. The causes may be dietary, due to reduced fibre and fluid intake, certain medications or intestinal obstruction.

Diabetes insipidus

This is a rare metabolic disorder in which a person produces large quantities of dilute urine and is constantly thirsty. It is due to the deficiency of the hormone ADH which regulates reabsorption of water in the kidneys. It is treated by administration of the hormone.

Diabetes mellitus

This is a carbohydrate metabolism disorder in which sugars are not oxidised to produce enough energy due to lack of the pancreatic hormone insulin. The accumulation of sugar leads to its appearance in the blood, then in the urine. Symptoms of diabetes mellitus include thirst, loss of weight and excessive production of urine.

Diarrhoea

This condition presents with frequent bowel evacuation or the passage of abnormally soft or liquid faeces. It may be caused by intestinal infections or other forms of intestinal inflammation such as colitis or irritable bowel syndrome.

Gallstones

This is a hard, pebblelike mass which is formed within the gall bladder. The condition may be asymptomatic or indigestion and colicky pain may be present. Changes in the composition of bile cause cholesterol and/or the bile pigment bilirubin to form stones. Stagnation of bile and inflammation of the gall bladder increase the concentration of bile and promote stone formation.

Haemorrhoids

This condition presents with abnormal dilatation of veins in the rectum. It is caused by increased pressure in the venous network of the rectum. If the haemorrhoids are chronic they may be seen or felt as soft swellings in the anus.

Heartburn

This is a burning sensation felt behind the sternum and often appears to rise from the abdomen, up the oesophagus, towards or into the throat. It is caused by regurgitation of the acidic stomach contents.

Hepatitis

This is an inflammation of the liver caused by viruses, toxic substances or immunological abnormalities.

■ **Hepatitis A** – this is highly contagious and is transmitted by the faecal/oral route. It is transmitted by ingestion of contaminated food, water or milk. The incubation period is 15 to 45 days.

■ **Hepatitis B** – this is also known as serum hepatitis and is more serious than Hepatitis A. It lasts longer and can lead to cirrhosis, cancer of the liver and a carrier state. It has a long incubation period of one and a half to two months. The symptoms may last from weeks to months. The virus is usually transmitted through infected blood, serum or plasma. However, it can spread by oral or sexual contact as it is present in most body secretions.

■ **Hepatitis C** – this can cause acute or chronic hepatitis and can also lead to a carrier state and liver cancer. It is transmitted through blood transfusions or exposure to blood products. Most clients with hepatitis are jaundiced but they can appear to be entirely healthy. Hepatitis as a side effect of drugs and alcohol intake is not infective.

Hernia

This is an abnormal protrusion of an organ or part of an organ through the wall of the body cavity in which it normally lies.

Hiatus hernia

This is the most common type of hernia and occurs when part of the stomach is protruding into the chest. This sometimes causes no symptoms at all, but it can cause acid reflux when acid from the stomach passes to the oesophagus, causing pain and heartburn.

Jaundice

This is a yellowing of the skin or whites of the eyes caused by excessive bilirubin (bile pigment) in the blood. It is caused by a malfunctioning gall bladder or obstructed bile duct.

Irritable bowel syndrome

This is a common condition in which there is recurrent abdominal pain with constipation and/or diarrhoea and bloating. Clients with stress and hectic lifestyles are more vulnerable to this illness. They usually defecate infrequently, usually in the morning, but may feel that their bowel is not empty or they may pass stool-like pellets.

Stress

Stress can be defined as any factor that affects physical or emotional well-being. Signs of stress affecting the digestive system include the development of ulcers, irritable bowel syndrome and indigestion.

Ulcers

This is a break in the skin or a break in the lining of the alimentary tract which fails to heal and is accompanied by inflammation. Peptic, duodenal and gastric ulcers can present with increased acidity, epigastric pain and heartburn. This may be worst when hungry or after consumption of irritating foods and alcohol, such as spicy or fatty foods, mayonnaise, wines and spirits. It can present with similar symptoms to a hiatus hernia and reflux.

Interrelationships with other systems

The digestive system links to the following body systems:

Cells and tissues

In areas of the digestive system, such as the small intestine, where absorption of nutrients is required, there is a thin lining of simple epithelium to allow for speedy absorption.

Skin

One of the skin's functions is in vitamin D production which helps in the absorption of calcium in the small intestine.

Skeletal

The maxilla and mandible, the larger bones in the face, support the jaw and teeth when food is ingested in the mouth.

Muscular

The action of peristalsis is due to the involuntary contraction of the smooth muscle in the alimentary canal that propels the food through the digestive tract. Skeletal facial muscles, such as the masseter and buccinator, assist in chewing.

Circulatory

Nutrients are carried in the body to nourish the cells and tissues, and waste products are carried away by the blood to be eliminated.

Lymphatic

Lymphatic vessels called lacteals (in the villi of the small intestine) assist digestion by absorbing the products of fat digestion.

Respiratory

Oxygen absorbed from the lungs activates glycogen from the digestive system to produce energy for cell metabolism.

Nervous

All the organs of the digestive system are stimulated by nerve impulses.

Endocrine

The pancreas secretes insulin from cells called the islets of Langerhans which help to control blood sugar levels.

Key words associated with the digestive system

digestion	chyme	lacteal
ingestion	small intestine	fatty acids
absorption	duodenum	glycerol
assimilation	jejenum	liver
elimination	ileum	hepatic portal vein
alimentary tract	villi	large intestine
peristalsis	gall bladder	caecum
mouth	bile	appendix
pharynx	pancreas	colon
oesophagus	pancreatic juice	ascending colon
saliva	intestinal juice	transverse colon
salivary amylase	emulsification	descending colon
starch	trypsin	sigmoid colon
stomach	pancreatic amylase	rectum
gastric juice	pancreatic lipase	faeces
enzyme	glucose	defecation
pepsin	peptidases	anus
hydrochloric acid	polypeptides	
mucus	amino acids	

Summary of the digestive system

- **Digestion** is the process of breaking down food and involves **ingestion, mechanical** and **chemical digestion, absorption, assimilation** and **elimination.**

- Digestion occurs in the **alimentary tract,** which extends from the mouth to the anus.

- **Peristalsis** is the coordinated, rhythmical contraction of the muscles in the wall of the alimentary tract.

- The digestive system consists of the **mouth, pharynx, oesophagus, stomach, small intestine, large intestine** and **anus.**

- The accessory organs to digestion are the **pancreas, gall bladder** and **liver.**

- The digestive system commences in the **mouth,** where food is broken down by mastication and mixed with **saliva.**

- **Saliva** contains the enzyme **salivary amylase** which commences the digestion of **starch** in the mouth.

- The muscles of the **pharynx** force the food down the **oesophagus** to the **stomach.**

- In the **stomach** food is mixed with **gastric juice** containing the enzyme **pepsin,** which starts the breakdown of proteins,

hydrochloric acid, to kill germs present in food and to prepare it for intestinal digestion, and **mucus**, which protects the stomach lining from the damaging effects of the acidic gastric juice.

■ The functions of the **stomach** are to churn and break up large particles of food mechanically, mix food with gastric juice to begin the chemical breakdown of food, commence the digestion of protein and absorb alcohol.

■ Food stays in the stomach for approximately five hours until it has been churned down to a liquid state called **chyme**.

■ **Chyme** is then released at intervals into the first part of the **small intestine.**

■ The **small intestine** consists of three parts – **duodenum, jejunum** and **ileum** (where absorption of food mainly takes place).

■ Special features of the small intestine are the thousands of minute projections called **villi**, a network of capillaries, into which the nutrients pass to be absorbed into the bloodstream.

■ The **small intestine** continues the mechanical breakdown of food by **peristalsis**, while the chemical digestion is brought about by **bile** (released by the gall bladder), enzymes in **pancreatic juice** (released by the pancreas) and **intestinal juice** (released by the walls of the small intestine which prepare the food to be absorbed into the bloodstream).

■ The function of **bile** is to neutralise the **chyme** and break up any fat droplets by **emulsification**.

■ The enzymes contained within **pancreatic juice** continue the digestion of protein (**trypsin**), carbohydrates (**pancreatic amylase**) and fats (**pancreatic lipase**).

■ **Intestinal juice** is released by the glands of the small intestine and completes the final breakdown of nutrients, including **simple sugars** to **glucose** and **protein** to **amino acids.**

■ Protein digestion is completed by **peptidases**, which split short-chain **polypeptides** into **amino acids.**

■ The absorption of the digested food takes place in the **jejunum** and mainly in the **ileum**.

■ Each villus contains a lymph vessel called a **lacteal** into which fatty acids and glycerol can pass.

■ Simple sugars from carbohydrate digestion and amino acids from protein digestion pass into the bloodstream via the **villi** and are then carried to the **liver**, via the **hepatic portal vein**, to be processed.

■ Products of fat digestion pass into the **lacteals** (intestinal lymphatics) which absorb the fat molecules and carry them through the lymphatic system before they reach the blood circulation.

■ Vitamins and minerals travel across to the blood capillaries of the villi and are absorbed into the bloodstream to assist in normal body functioning and cell metabolism.

■ **Glucose**, the end product of carbohydrate digestion, is used to provide energy for the cells to function.

■ **Amino acids,** the end products of protein digestion, are used to produce new tissues, repair damaged cell parts and formulate enzymes, plasma proteins and hormones.

■ **Fatty acids** and **glycerol** are the end products of fat digestion.

- Fats are used primarily to provide heat and energy in addition to glucose. Those fats which are not required immediately by the body are used to build cell membranes and some are stored under the skin or around vital organs such as the kidneys and the heart.

- When all the body's nutrients have been assimilated by the body, the undigested food is passed into the large intestine where it is eventually eliminated from the body.

- The **large intestine** is made up of the **caecum**, **appendix**, **colon** and **rectum**.

- The **colon** is the main part of the large intestine and is divided into **ascending**, **transverse**, **descending** and **sigmoid** colons.

- The **rectum** is the last part of the large intestine, where **faeces** are stored before **defecation**.

- The functions of the large intestine are the absorption of most of the water from the faeces, formation and storage of faeces, production of mucus to lubricate the passage of faeces and the expulsion of faeces from the body.

- The **anus** is an opening at the lower end of the alimentary canal (the anal canal), through which faeces are discharged.

- The **liver** is the largest gland in the body and is an accessory organ to digestion with many metabolic functions.

- The functions of the **liver** include the secretion of bile, regulation of blood sugar levels, regulation of amino acid levels, regulation of the fat content of blood, regulation of plasma proteins, detoxification, storage and the production of heat.

- The **pancreas** is also an accessory organ to digestion. Its exocrine function is the secretion of pancreatic juice.

- The **gall bladder** is attached to the posterior and inferior surface of the liver and its function is to store bile produced by the liver until it is needed.

Multiple-choice questions

1. **The alimentary tract is a long, continuous, muscular tube extending from the:**
    a   mouth to anus
    b   stomach to anus
    c   small intestine to anus
    d   large intestine to anus

2. **Which of the following completes digestion?**
    a   stomach
    b   gall bladder
    c   small intestine
    d   large intestine

3. **Which of the following statements is true?**
    a   the liver is situated in the upper left-hand side of the abdominal cavity
    b   the liver's internal structure is made up of cells called hepatocytes
    c   bile is stored in the liver and released by the pancreas
    d   when blood sugar levels are low the liver cells store excess glucose

4. **Which of the following is produced in the stomach?**
    a   bile
    b   pancreatic juice
    c   pepsin
    d   maltase

5. **The commencement of protein digestion occurs in the:**
    a   mouth
    b   small intestine
    c   stomach
    d   pancreas

6. **Which of the following is responsible for the chemical reactions of digestion?**
    a   homeostasis
    b   enzymes
    c   absorption
    d   peristalsis

7. **Salivary amylase commences:**
    a   carbohydrate digestion
    b   protein digestion
    c   fat digestion
    d   vitamin and mineral digestion

8. **The main constituents of gastric juice are:**
    a   gastrin and pepsin
    b   gastrin and pepsinogen
    c   pepsin, hydrochloric acid and mucus
    d   gastric amylase

9. **Trypsin is an enzyme produced by the:**
    a   duodenum
    b   liver
    c   pancreas
    d   gall bladder

10. **Where does peristalsis occur?**
    a   only in the mouth
    b   only in the small intestine
    c   only in the stomach
    d   in all sections of the alimentary canal

11. **Vitamins and minerals are absorbed into the bloodstream:**
    a   via the liver cells
    b   via the villi in the small intestine
    c   via the lacteals in the small intestine
    d   via the hepatic portal vein

12. The main part of the large intestine is the:

a duodenum
b colon
c caecum
d ileum

13. The three sections of the small intestine from beginning to end are:

a ascending, transverse and descending
b duodenum, jejenum and ileum
c jejunum, ileum and duodenum
d duodenum, ileum and jejenum

14. The colon primarily absorbs which substance?

a carbohydrates
b proteins
c fats and lipids
d water

the urinary system

Introduction

The kidneys and their associated structures are all part of the excretory system, along with the skin, lungs and intestines, which also contribute to the job of waste elimination in the body.

The urinary system is made up the kidneys, ureters, bladder and urethra which are involved in the processing and elimination of normal metabolic waste from the body and can be likened to the body's 'plumbing system'.

Objectives

By the end of this chapter you will be able to recall and understand the following knowledge:

■ the functions of the urinary system
■ the structure and functions of the individual parts of the urinary system (kidneys, ureters, urinary bladder and urethra)
■ the interrelationships between the urinary and other body systems
■ disorders of the urinary system.

Functions of the urinary system

The prime function of the urinary system is to help maintain homeostasis by controlling the composition, volume and pressure of blood. It does this by removing and restoring selected amounts of water and dissolved substances.

Waste products, such as urea and uric acid, along with excess water and mineral salts, must be removed from the body in order to maintain good health. If these waste materials were allowed to accumulate in the body they would cause ill health. The primary function of the urinary system, therefore, is to regulate the composition and the volume of body fluids in order to provide a constant internal environment for the body.

Structures of the urinary system

The urinary system consists of the following parts:

- two **kidneys** which secrete urine

- two **ureters** which transport urine from the kidneys to the bladder

- one **urinary bladder** where urine collects and is stored temporarily

- one **urethra** through which urine is discharged from the bladder and out of the body.

Kidneys

The kidneys are bean-shaped organs lying on the posterior wall of the abdomen on either side of the spine, between the level of the twelfth thoracic vertebra and the third lumbar vertebra. Due to the presence of the liver, the kidney on the right side of the body is slightly higher than the one on the left.

Structure of the kidney

A kidney has an outer fibrous renal capsule and is supported by adipose tissue. It has two main parts:

- **Outer cortex** – this is reddish-brown and is the part where fluid is filtered from blood.

- **Inner medulla** – this is paler in colour and is made up of conical-shaped sections called **renal pyramids**. This is the area where some materials are selectively reabsorbed into the bloodstream.

There is a large area in the centre of the kidney called the **renal pelvis**, which is a funnel-shaped cavity that collects urine from the renal pyramids in the medulla and drains it into the ureter. The medial border of the kidney is called the **hilus** and is the area where the renal blood vessels leave and enter the kidney.

Nephron

The cortex and the medulla contain tiny blood-filtration units called nephrons. Nephrons are the functional units of the kidney and they extend from the renal capsule through the cortex and medulla to the cup-shaped renal pelvis. Nephrons are approximately 2 to 4 cm long and a single kidney has more than a million nephrons.

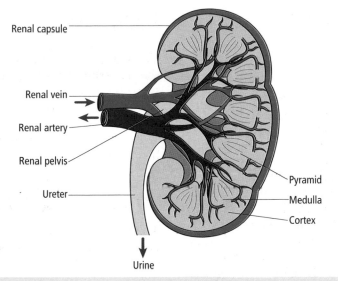

Figure 12.1   Structure of a kidney

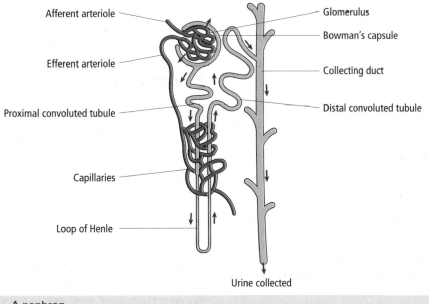

Figure 12.2   A nephron

Urine production

Urine is produced by three processes:

- **filtration**
- **selective reabsorption**
- **collection.**

Filtration

The blood that needs to be processed enters the medulla of the kidney from the renal artery. Inside the kidney the renal artery splits into a network of capillaries called the glomerulus which filter the waste. Almost encasing the glomerulus lies a sac called the Bowman's capsule.

The blood pressure in the glomerulus is maintained at a high level, assisted by the fact that the arteriole feeding into the glomerulus has a larger diameter than the arteriole leaving it. This pressure forces fluid out through the walls of the glomerulus, together with some of the substances of small molecular size able to pass through the capillary walls into the Bowman's capsule. This process constitutes simple filtration.

Selective reabsorption

The filtered liquid continues through a series of twisted tubes called the convoluted tubules, which are surrounded by capillaries. The tubules of the nephron that lead away from the Bowman's capsule are known as the proximal convoluted tubules and they straighten out to form a long loop called the loop of Henle. There are then another series of twists called the distal convoluted tubules, which lead to a straightened collecting duct, which leads to the pelvis of the kidney and on to the ureter.

The composition of the filtered liquid alters as it flows through the convoluted tubules. Some substances contained within the waste, such as glucose, amino acids, mineral salts and vitamins, are reabsorbed into the bloodstream as the body cannot afford to lose them.

This reabsorption process is selective as the amount of these substances which passes back in the bloodstream depends on the level already present in the bloodstream and within the body. The reabsorption of salts and water is variable and is associated with the maintenance of a stable condition of acidity/alkalinity and electrolyte (sodium and potassium) balance of body fluids. Excess water, salts and the waste product urea are all filtered and processed through the kidneys and the treated blood leaves the kidney via the renal vein.

Some substances are not removed from the blood completely in the glomerular filtrate, such as the residue of medicinal drugs. These substances are passed from the blood, in particular, into the distal convoluted tubule by secretion in order that they may be excreted in the urine.

Collection

The wastes remaining in the distal convoluted tubule (now known as urine) then flow on via a collecting tubule to the renal pelvis of the kidney. From here it passes into the ureter to be passed to the bladder and urethra to be excreted.

Composition of urine

Urine is the concentrated filtrate from the kidneys. Its composition is 96 per cent water, 2 per cent urea and 2 per cent other substances, such as uric acid, creatinine, sodium, potassium, phosphates, chlorides, sulphates, excess vitamins and drug residues.

Urine is a pale, watery fluid, varying in colour according to its composition and quantity. Urine is usually acidic and its pH varies between 4.5 and 7.4 depending on the blood pH.

The salts, chiefly sodium chloride, must be reabsorbed in the kidney tubules or at least removed in sufficient quantities necessary to keep the blood at its normal pH (7.4) and to maintain the water and electrolyte balance. As the pH and salt concentration are both essential to the life of the blood and tissue cells, the functions of the kidneys are of paramount importance.

How much you have of certain substances in your urine is a good indication of the state of health of your body. Urine tests are often used to diagnose a disorder. If urine contains glucose it could indicate a person has diabetes. Protein in the urine could indicate that the kidneys are failing. Urine can be used to confirm a pregnancy. A fertilised ovum releases a hormone which the mother excretes in the urine. This can usually be detected from about 14 days after fertilisation and is the basis of most pregnancy tests.

Functions of the kidney

The functions of the kidney are:

- filtration of impurities and metabolic waste from blood and preventing poisons from fatally accumulating in the body
- regulation of water and salt balance in the body
- maintenance of the normal pH balance of blood
- formation of urine
- regulation of blood pressure and blood volume.

Role of the kidneys in fluid balance

The amount of fluid taken into the body must equal the amount of fluid excreted from it in order for the body to maintain a constant internal environment. The balance between water intake and water output is controlled by the kidneys.

Water intake

Water is mainly taken into the body as liquid through the process of digestion. However, some is also released through the cells' metabolic activities.

Water output

Water is lost from the body in the following ways:

- through the kidneys as urine
- through the alimentary tract as faeces
- through the skin as sweat
- through the lungs as saturated, exhaled breath.

The kidneys are responsible for regulating the amount of water contained within the blood. The amount of water reabsorbed into the blood is controlled by the antidiuretic hormone (ADH) which is stored and released into the blood by the posterior lobe of the pituitary gland.

The release of ADH is triggered by dehydration. The hypothalamus detects when the water concentration of blood is low and triggers the release of ADH. An increase in the level of ADH increases the amount of water that is reabsorbed from the nephron into the blood.

The reabsorption of water from the nephron into the blood decreases the volume of urine expelled from the kidneys and increases the hydration level of the blood. This mechanism reduces the amount of water in the blood to an acceptable level.

Key note

This important negative feedback mechanism between the nervous and endocrine systems maintains the blood concentration within normal limits and is the means by which fluid balance is controlled in the body.

Factors affecting fluid balance

Factors affecting fluid balance in the body include:

- **Body temperature** – if the body temperature increases more water is lost from the body in sweat.

- **Diet** – a high salt intake can result in increased water reabsorption which reduces the volume of urine produced. Diuretics, such as alcohol, tea and coffee, can also increase the volume of urine.

- **Emotions** – nervousness can result in an increased production of urine.

- **Blood pressure** – when the blood pressure inside the kidney tubules rises, less water is reabsorbed and the volume of urine will be increased (but not decreased). When the blood pressure inside the kidney tubules falls, more water is reabsorbed into the blood and the volume of urine will be decreased.

Ureters

The ureters are two very fine, muscular tubes which transport urine from the renal pelvis of the kidney to the urinary bladder. They consist of three layers of tissue:

- an outer layer of fibrous tissue
- a middle layer of smooth muscles
- an inner layer of mucous membrane.

Function of the ureters

Their function is to propel urine from the kidneys into the bladder by the peristaltic contraction of their muscular walls.

Urinary bladder

This is a pear-shaped sac which lies in the pelvic cavity, behind the symphysis pubis. The size of the bladder varies according to the amount of urine it contains. The bladder is composed of four layers of tissue:

- a serous membrane which covers the upper surface
- a layer of smooth muscular fibres
- a layer of adipose tissue
- an inner lining of mucous membrane.

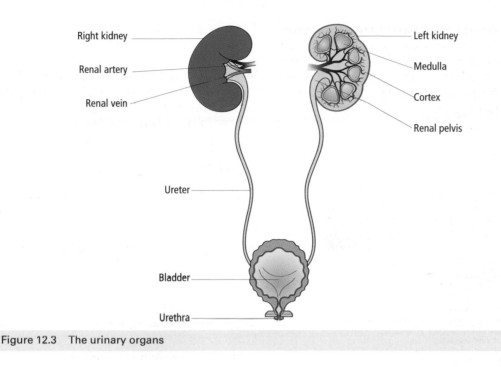

Right kidney

Renal artery

Renal vein

Left kidney

Medulla

Cortex

Renal pelvis

Ureter

Bladder

Urethra

Figure 12.3   The urinary organs

Functions of the urinary bladder

The urinary bladder stores urine. It expels urine from the body, assisted by the muscular wall of the bladder, the lowering of the diaphragm and the contraction of the abdominal cavity.

The expelling of urine from the bladder is called micturition and is a reflex over which there is voluntary control. When the volume of urine in the bladder causes it to expand, stretch receptors in the bladder wall are stimulated to trigger urination. The micturition reflex causes the detrusor muscle in the wall of the bladder to contract and the internal urethral sphincter to relax. It is the combination of both the micturition reflex and voluntary relaxation of the urethral sphincter that allows urination to occur.

Urethra

This is a canal which extends from the neck of the bladder to the outside of the body. The length of the urethra differs in males and females. The female urethra is approximately 4 cm in length, whereas the male urethra is longer at approximately 18 to 20 cm. The exit from the bladder is guarded by a round sphincter of muscles which must relax before urine can be expelled from the body.

The urethra is composed of three layers of tissue:

- a muscular coat continuous with that of the bladder
- a thin, spongy coat which contains a large number of blood vessels
- a lining of mucous membrane.

Function of the urethra

The urethra serves as a tube through which urine is discharged from the bladder to the exterior. The urethra is longer in a male and also serves as a conducting channel for semen.

Disorders of the urinary system

Cancer of the bladder

This usually presents with blood in the urine and urgency and pain on passing urine. Secondary symptoms may arise if it has spread to the lungs, liver, lymph nodes and neighbouring tissues.

Cystitis

This is an inflammation of the urinary bladder, usually caused by infection of the bladder lining. Common symptoms are pain just above the pubic bone, lower back or inner thigh, blood in the urine and frequent, urgent and painful urination with a burning sensation. This condition is very common in women due to the shorter length of the female urethra.

Incontinence

This is a condition in which the individual is unable to control urination voluntarily. Loss of muscle tone and problems with innervation are associated with this condition.

Kidney stones

These are deposits of substances found in the urine which form solid stones in the renal pelvis of the kidney, ureter or bladder. This condition can be extremely painful. Stones are usually removed by surgery.

Nephritis

A general, non-specific term used to describe inflammation of the kidney. Glomerulonephritis (also known as Bright's disease) is an inflammation of the glomeruli in the kidneys. This condition is characterised by blood in the urine, fluid retention and hypertension.

Pyelonephritis

This is a bacterial infection of the kidney. In acute pyelonephritis there is pain in the back, high temperature and shivering fits. Treatment is usually with antibiotics.

Urinary tract infection

This is a bacterial infection of one or more of the structures of the urinary system. Symptoms include fever, lower back pain, frequency of urination, a burning sensation on passing urine (urine may be bloodstained and cloudy). If the infection is severe there may be blood and pus in the urine.

Interrelationships with other systems

The urinary system links to the following body systems:

Cells and tissues
Transitional epithelium lines urinary organs, such as the bladder, which change shape when stretched.

Skin
Like the urinary system the skin is also an excretory organ. When the skin loses excess water through sweating, the kidneys release less water in the urine to help maintain the body's fluid balance.

Skeletal
The kidneys and the bones of the skeleton help to control the amount of calcium in the blood by storing some in the bones and excreting some from the body in urine.

Muscular
Smooth muscle is responsible for the passage of urine through the urinary tract.

Circulatory
The kidneys filter the blood to avoid poisons from fatally accumulating in the body.

Nervous
The relaxation and contraction of the bladder and closing and opening of the sphincter muscles is under the control of the autonomic nervous system (sympathetic and parasympathetic nervous systems).

Digestive
Water is an essential nutrient which is needed by every part of the body and is taken in through the process of digestion to aid the metabolic process. The colon absorbs most of the water from the faeces in order to conserve moisture in the body.

Key words associated with the urinary system

kidney	glomerulus	collecting duct
renal artery	Bowman's capsule	renal pelvis
renal vein	proximal convoluted tubule	urine
cortex		ureter
medulla	loop of Henle	urinary bladder
nephron	distal convoluted tubule	urethra

Summary of the urinary system

- The organs that contribute to the elimination of waste in the body are the kidneys, lungs, skin and the digestive system.

- The organs of the urinary system are the **kidneys, ureters, urinary bladder** and **urethra.**

- The **kidneys** are bean-shaped organs, lying on the posterior wall of the abdomen.

- The **kidney** has two main parts – the outer **cortex** where fluid is filtered from blood, and the inner **medulla** which is the area where some materials are selectively reabsorbed into the bloodstream.

- The **cortex** and the **medulla** contain tiny blood-filtration units called **nephrons.**

- Urine is produced by three processes – **filtration, selective reabsorption** and **collection.**

- Blood to be processed enters the kidneys via the **renal artery.**

- **Filtration** takes place inside a network of capillaries in the **nephron** called the **glomerulus.**

- The sac encasing the **glomerulus** is called the **Bowman's capsule.**

- The filtered liquid then continues through a series of twisted tubes called the **convoluted tubules**, to the **loop of Henle** and the **distal convoluted tubule,** before passing to the **collecting duct** and the **renal pelvis.**

- The composition of the filtered liquid alters as it flows through the **convoluted tubules.**

- Some substances in the filtrate, like glucose, amino acids, mineral salts and vitamins, are reabsorbed into the bloodstream via the **renal vein.**

- From the **distal convoluted tubule** the filtrate then flows into the **collecting duct** (as urine) and passes to the pelvis of the kidney to be passed to the **ureter** and **bladder.**

- The composition of urine is 96 per cent water, 2 per cent urea and 2 per cent other substances (uric acid, creatinine, sodium, potassium, phosphates, chlorides, sulphates, excess vitamins and drug residues).

- Functions of the kidneys include filtration of impurities and metabolic waste from blood, regulation of water and salt balance, formation of urine and regulation of blood pressure and volume.

- The **ureters** are muscular tubes that transport urine from the pelvis of the kidney to the urinary bladder.

- The **urinary bladder** is a pear-shaped sac which lies in the pelvic cavity, behind the symphysis pubis.

- The urinary bladder functions as a storage organ for urine.

- The **urethra** is a canal which extends from the neck of the bladder to the outside of the body.

- The urethra serves as a tube through which urine is discharged from the bladder to the exterior and as a conducting channel for semen in men.

Multiple-choice questions

1. **The function of the kidneys is the:**
   a   filtering of impurities from the blood
   b   regulation of water and salt balance
   c   formation of urine
   d   all of the above

2. **Which of the following is not considered an excretory organ?**
   a   digestive
   b   skin
   c   muscular
   d   respiratory

3. **Blood is filtered inside what section of the kidney?**
   a   glomerulus
   b   Bowman's capsule
   c   loop of Henle
   d   proximal convoluted tubule

4. **The blood-filtration unit inside a kidney is known as:**
   a   hilus
   b   renal pyramid
   c   nephron
   d   medulla

5. **Which of the following best describes the position of the kidneys?**
   a   posterior of abdomen between the level of twelfth thoracic and fifth lumbar vertebrae
   b   posterior of thorax, between the level of twelfth thoracic and fifth lumbar vertebrae
   c   posterior of abdomen between the level of twelfth thoracic and third lumbar vertebrae
   d   posterior of thorax, between the level of twelfth thoracic and third lumbar vertebrae

6. **Which of the following statements is true?**
   a   filtered blood leaves the kidney via the renal artery
   b   excess water, salts and urea are all filtered and processed through the kidneys
   c   blood to be processed enters the medulla from the renal vein
   d   the renal artery splits into a network of capillaries called the Bowman's capsule

7. **The hormone responsible for controlling water reabsorption in the kidneys is:**
   a   insulin
   b   antidiuretic hormone (ADH)
   c   oxytocin
   d   adrenocorticotrophic hormone

8. **The function of the ureter is to:**
   a   propel urine from the bladder to the exterior
   b   store urine
   c   filter impurities
   d   propel urine from the kidneys to the bladder

9. **The bladder is situated:**
   a   in the abdominal cavity behind the intestines
   b   in the pelvic cavity behind the symphysis pubis
   c   on the posterior of the abdominal cavity
   d   behind the urethra

10. **The condition cystitis commonly affects females as:**
    a   women tend to have a weaker bladder than men
    b   women have a shorter urethra
    c   women have shorter ureters
    d   women have smaller kidneys

11. **Which of the following factors affects fluid balance?**
   a   diet
   b   body temperature
   c   blood pressure
   d   all of the above

12. **Which of the following also serves as a conducting channel for semen in the male?**
   a   ureter
   b   urethra
   c   bladder
   d   none of the above

web resource links

http://www.innerbody.com/

http://bubl.ac.uk/link/h/humananatomy.htm

http://www.bartleby.com/107/

http://www.kensbiorefs.com/humphy.html

http://www.getbodysmart.com/

http://www.msjensen.gen.umn.edu/webanatomy/

http://science.nhmccd.edu/biol/ap1int.htm

http://science.nhmccd.edu/biol/ap2int.htm

http://www.meddean.luc.edu/lumen/MedEd/GrossAnatomy/dissector/mml/index.htm

http://www.nlm.nih.gov/medlineplus/anatomy.html

http://www.eskeletons.org/

http://webschoolsolutions.com/patts/systems/anatomy.htm

http://www.rad.washington.edu/atlas/

http://arbl.cvmbs.colostate.edu/hbooks/

http://www.meddean.luc.edu/lumen/MedEd/GrossAnatomy/cross_section/index.html

http://www.vh.org/adult/provider/anatomy/HumanAnatomy/CrossSectionAtlas.html

glossary

glossary

Abdominal nodes: lymph nodes located within the abdominal cavity along the branches of the abdominal aorta. They drain lymph from organs within the abdominal cavity.

Abducens nerve: mixed nerve that innervates only the lateral rectus muscle of the eye.

Abduction: movement of a limb away from the midline.

Absorption: movement of soluble materials out through the walls of the small intestine to be delivered to various parts of the body.

Accessory nerve: functions primarily as a motor nerve, innervating muscles in the neck and upper back, as well as muscles of the palate, pharynx and larynx.

Acetylcholine: neurotransmitter substance which diffuses across the junction and stimulates the muscle fibres to contract.

Acne vulgaris: common inflammatory disorder of the sebaceous glands which leads to the overproduction of sebum. It involves the face, back and chest and is characterised by the presence of comedones, papules and, in more severe cases, cysts and scars.

Acromegaly: increase in the size of the hands, feet and face due to excessive production of the growth hormone.

Actin: protein found in muscle that plays an important role in muscle contraction.

Active transport: energy-dependent process in which certain substances with larger molecules are able to cross cell membranes against a concentration gradient.

Addison's disease: condition caused by undersecretion of corticosteroid hormones. Symptoms include loss of appetite, weight loss, brown pigmentation around joints, low blood sugar, low blood pressure, tiredness and muscular weakness.

Adduction: movement of a limb towards the midline.

Adductors: group of four muscles on the medial aspect of the thigh.

Adipose tissue: type of tissue containing fat cells, found in the subcutaneous layer of skin.

ADP (adenosine diphosphate): compound that releases the energy needed for processes such as muscle contraction.

Adrenal: one of two triangular-shaped glands which lie on top of each kidney. They consist of two parts, an outer cortex and an inner medulla.

Adrenaline: hormone secreted by the medulla of the adrenal glands. It prepares the body for 'fright, fight or flight' response (sympathetic nervous system), having widespread effects on the circulation, muscles and glucose metabolism.

Adrenocorticotrophic hormone (ACTH): secreted from the anterior lobe of the pituitary gland, it stimulates and controls the growth and hormonal output of the adrenal cortex.

Agonist/prime mover: main activating muscle.

AIDS (acquired immune deficiency syndrome): condition contracted as a result of the human immunodeficiency virus (HIV) which progressively destroys the immunity of the individual.

Albinism: inherited absence of pigmentation in the skin, hair and eyes, resulting in white hair, pink skin and eyes. The pink colour is produced by underlying blood vessels which are normally masked by pigment. Other clinical signs of this condition include poor eyesight and sensitivity to light.

Alimentary tract: long, continuous, muscular tube, extending from the mouth to the anus.

Allergic reaction: disorder in which the body becomes hypersensitive to a particular allergen. The body produces histamine in the skin, as part of the body's defence or immune system.

Alveoli: tiny air sacs inside the lungs.

Amenorrhea: absence or stopping of the menstrual period.

Amino acids: end products of protein digestion.

Anaemia: condition where the haemoglobin level in the blood is below normal.

Anagen: active, growing phase of hair.

Androgens: collective term for male hormones.

Aneurysm: abnormal, balloonlike swelling in the wall of an artery.

Angina: pain in the left side of the chest and usually radiating to the left arm. It is caused by insufficient blood to the heart muscle and usually on exertion or excitement.

Ankylosing spondylitis: systemic joint disease characterised by inflammation of the intervertebral disc spaces, costo-vertebral and sacroiliac joints.

Anorexia nervosa: psychological illness in which people starve themselves or use other techniques, such as vomiting or laxatives, to induce weight loss.

Antagonists: two muscles or sets of muscles which pull in opposite directions to each other.

Antibody: specific protein produced to destroy or suppress antigens.

Antidiuretic hormone (ADH): hormone secreted from the posterior lobe of the pituitary. It increases water reabsorption in the renal tubules of the kidneys.

Antigen: any substance that the body regards as foreign or potentially dangerous and against which it produces an antibody.

Anus: opening at the lower end of the alimentary tract through which faeces are discharged.

Anxiety: psychological condition which can be defined as fear of the unknown. As an illness it can vary from a mild form to panic attacks and severe phobias that can be disabling socially, psychologically and, at times, physically.

Aorta: main artery of the systemic circulation.

Apocrine gland: type of sweat gland found in the genital and underarm regions.

Appendicitis: acute inflammation of the appendix.

Appendicular skeleton: part of skeleton consisting of the shoulder girdle, bones of the upper and lower limbs and bones of the pelvic girdle.

Appendix: short, thin, blind-end tube that is attached to the end of the caecum and has no known function in humans.

Arteriosclerosis: circulatory disorder characterised by a thickening, narrowing, hardening and loss of elasticity of the walls of the arteries.

Artery: type of blood vessel with thick, muscular and elastic walls. It carries blood away from the heart.

Arthritis – gout: joint disorder due to deposition of excessive uric acid crystals accumulating in the joint cavity.

Arthritis – osteoarthritis: joint disease characterised by the breakdown of articular cartilage, growth of bony spikes, swelling of the surrounding synovial membrane and stiffness and tenderness of the joints.

Arthritis – rheumatoid: chronic inflammation of peripheral joints, resulting in pain, stiffness and potential damage to joints.

Assimilation: process by which digested food is used by the tissues after absorption.

Asthma: attacks of shortness of breath and difficulty in breathing due to spasm or swelling of the bronchial tubes.

Atony: state in which the muscles are floppy and lacking their normal degree of elasticity.

ATP (adenosine triphosphate): compound that stores the energy needed for processes such as muscle contraction.

Atrium: upper chamber of heart.

Atrophy: wasting of muscle tissue due to undernourishment or lack of use.

Autonomic nervous system: part of the nervous system which supplies impulses to smooth muscles, cardiac muscle, skin, special senses and proprioceptors. It consists of a sympathetic and parasympathetic division.

Axial skeleton: forms the main axis or central core of the body and consists of the skull, vertebral column, sternum and ribs.

Axillary nodes: lymph nodes located in the underarm region. They drain lymph from the upper limbs, wall of the thorax, breasts and upper wall of the abdomen.

Axon: long, single nerve fibre extending from the cell body. Its function is to transmit impulses away from the cell body.

Ball-and-socket joint: type of synovial joint formed when the rounded head of one bone fits into a cup-shaped cavity of another bone. Movement is possible in several directions.

Basal cell layer (stratum germinativum): deepest and innermost of the five layers of epidermis.

Basilar artery: artery in the base of the brain, formed by the joining together of the two vertebral arteries.

Bell's palsy: disorder of the seventh cranial nerve (facial nerve) that results in paralysis on one side of the face.

Benign: term used to describe a tumour that does not invade and destroy the tissue in which it originates or spread to distant sites in the body.

Biceps: muscle on the anterior of the upper arm.

Bile: green, alkaline liquid, produced in the liver and stored in the gall bladder.

Blood: fluid tissue or medium in which all materials are transported to and from individual cells in the body.

Blood pressure: amount of pressure exerted by blood on an arterial wall due to the contraction of the left ventricle.

Boil: begins as a small, inflamed nodule which forms a pocket of bacteria around the base of a hair follicle or a break in the skin.

Bone: hardest type of connective tissue in the body.

Bowman's capsule: cup-shaped end of a nephron and site of primary filtration of the blood into the kidney tubule.

Brachialis: muscle attaching to the distal half of the anterior surface of the humerus at one end and the ulna at the other.

Brachial plexuses: spinal nerves at the top of the shoulder supplying the skin and muscles of the arm, shoulder and upper chest.

Brachioradialis: anterior muscle of the forearm, connecting the humerus to the radius.

Brain: highly developed mass of nervous tissue that forms the upper part of the central nervous system.

Brain stem: enlarged extension upwards within the skull of the spinal cord, consisting of the medulla oblongata, the pons and the midbrain.

Breast/mammary gland: accessory female sex gland that produces milk following childbirth.

Bronchi: two short tubes which lead to and carry air into each lung.

Bronchiole: subdivision of the bronchial tree.

Bronchitis: chronic or acute inflammation of the bronchial tubes.

Buccinator: main muscle of the cheek. It is attached to both the upper and lower jaw.

Bulimia: psychological illness which is characterised by overeating (bingeing), followed by self-induced vomiting.

Bunion: swelling of the joint between the big toe and the first metatarsal.

Bursitis: inflammation of a bursa (small sac of fibrous tissue that is lined with synovial membrane and filled with synovial fluid).

Caecum: small pouch to which the appendix is attached and into which the ileum opens.

Calcaneum: bone at the end of the metatarsals.

Calcitonin: hormone secreted by the thyroid gland which controls the level of calcium in the blood.

Cancellous bone: open, spongy bone tissue found at the ends of long bones or at the centre of other bones.

Cancer: any malignant tumour. It arises from the abnormal and uncontrolled division of cells that invade and destroy the surrounding tissues.

Capillary: smallest blood vessels with a single layer of cells. It is responsible for supplying the cells and tissues with nutrients.

Carcinoma: malignant tumour that arises from epithelial cells.

Cardiac cycle: sequence of events between one heartbeat and the next, normally less than a second in duration.

Cardiac muscle: special type of involuntary muscle found only in the heart.

Carotid artery: either of the two main arteries of the neck whose branches supply the head and neck.

Carpal tunnel syndrome: condition characterised by pain and numbness in the thumb or hand, resulting from pressure on the median nerve of the wrist.

Carpals: eight small bones forming the wrist.

Cartilage: dense, connective tissue that consists of collagen and elastin fibres embedded in a strong, gel-like substance.

Cartilaginous joint: slightly movable joints which have a pad of fibrocartilage between the end of the bones making the joint.

Catagen: transitional stage of hair growth from active to resting.

Cell: basic unit of all living organisms.

Cell membrane: fine membrane that encloses the cell and protects its contents.

Cell regeneration: process by which the cells of the epidermis regenerate themselves by the process of mitosis.

Cell respiration: controlled exchange of nutrients for waste by the cell to activate the energy needed for the cell to function.

Central nervous system (CNS): part of the nervous system consisting of the brain and spinal cord.

Centrioles: small, spherical structures associated with cell division, contained within the centrosome.

Centromere: portion of a chromosome where the two chromatids are joined.

Centrosome: area of clear cytoplasm found next to the nucleus, containing the centrioles.

Cerebellum: cauliflower-shaped structure located at the posterior of the cranium and below the cerebrum.

Cerebral palsy: condition caused by damage to the central nervous system of a baby during pregnancy, delivery or soon after birth. The damage could be due to bleeding, lack of oxygen or other injuries to the brain.

Cerebrum: largest portion of the brain which makes up the front and top part of the brain.

Cervical plexuses: spinal nerves supplying the skin and muscles of the head, neck and upper region of the shoulders.

Cervical vertebrae: seven vertebrae of the neck.

Chloasma: pigmentation disorder which presents with irregular areas of increased pigmentation, usually on the face.

Cholesterol: fatlike material present in the blood and most tissues.

Chondrocyte: cartilage-forming cell.

Chromatid: pair of identical strands that are joined at the centromere and separate during cell division.

Chromatin: substance inside the nucleus that contains the genetic material.

Chromosome: threadlike structure in the cell nucleus that carries the genetic information.

Chyme: semi-liquid, acid mass in which food passes from the stomach to the small intestine.

Cilia: microscopic, hairlike processes on the exposed surfaces of certain epithelial tissue.

Circumduction: circular movement of a joint (360 degrees).

Cirrhosis of the liver: distorted or scarred liver as a result of chronic inflammation.

Clavicle: bone forming anterior of shoulder girdle.

Clear layer (stratum lucidum): epidermal layer below the most superficial layer (horny), consisting of small, tightly packed transparent cells which permit light to pass through.

Coccygeal plexus: supplies the skin in the area of the coccyx and the muscles of the pelvic floor.

Coccygeal vertebrae (coccyx): four fused vertebrae at the base of spine forming the tail bone.

Coccygeus: muscle forming the pelvic floor.

Colitis: inflammation of the colon.

Collagen: protein in the dermis which gives the skin its strength and resilience.

Collecting duct: vessel that leads to the pelvis of the kidney and on to the ureter.

Colon: main part of the large intestine, consisting of four sections – ascending, transverse, descending and sigmoid colons.

Comedone: collection of sebum, keratinised cells and wastes which accumulate in the entrance of a hair follicle. It may be open or closed.

Comminuted fracture: where a bone has splintered at the site of impact and smaller fragments of bone lie between the two main fragments.

Compact bone: hard portion of bone that makes up the main shaft of the long bones and the outer layer of other bones.

Complicated fracture: when a broken bone damages tissues and/or organs around it.

Compound fracture: open fracture where the broken ends of the bone protrude through the skin.

Concentric contraction: type of contraction when a muscle shortens to move the attachments closer.

Condyloid joint: type of synovial joint that is shaped so that the concave surface of one bone can slide over the convex surface of another bone in two directions.

Congenital heart disease: defect in the formation of the heart which usually decreases its efficiency.

Conjunctivitis: bacterial infection following irritation of the conjunctiva of the eye. The inner eyelid and eyeball appear red and sore and there may be a puslike discharge from the eye.

Connective tissue sheath: surrounds both the follicle and sebaceous gland. Its function is to supply the follicle with nerves and blood.

Constipation: difficulty in passing stools or infrequent evacuation of the bowel.

Contact dermatitis: inflammation of the skin caused by a primary irritant which causes the skin to become red, dry and inflamed.

Coracobrachialis: muscle that extends from the scapula to the middle of the humerus, along its medial surface.

Corrugator: muscle located on the inner edge of the eyebrow.

Cortex: middle layer of hair; outer, reddish-brown part of kidney where fluid is filtered from blood.

Cowper's glands: pair of small glands that open into the urethra at the base of the urethra. Their secretion contributes to the seminal fluid.

Cramp: prolonged, painful contraction of a muscle.

Cranial nerves: set of 12 pairs of nerves originating from the brain.

Cretinism: syndrome of dwarfism, mental retardation and coarseness of skin and facial features due to lack of thyroid hormones.

Crow's feet: fine lines around the eyes caused by habitual facial expressions and daily movement.

Cushing's syndrome: caused by the hypersecretion of the glucocorticoids. Symptoms include weight gain, reddening of the face and neck, excess growth of facial and body hair, raised blood pressure, loss of mineral from bone and sometimes mental disturbances.

Cuticle: fold of overlapping skin that surrounds the base of the nail, providing a protective seal against bacteria; outer layer of hair.

Cyst: abnormal sac containing liquid or a semi-solid substance.

Cystitis: inflammation of the urinary bladder, usually caused by infection of the bladder lining.

Cytoplasm: gel-like substance enclosed by the cell membrane.

Deep cervical nodes: lymph nodes located deep within the neck, which drain lymph from the larynx, oesophagus, posterior of the scalp and neck, and superficial part of chest and arm.

Defecation: elimination of faeces through the anal canal.

Deltoid: thick, triangular muscle that caps the top of the humerus and shoulder.

Dendrites: highly branched extensions of the nerve cell that receive and transmit stimuli towards the cell body.

Depression: psychological condition which combines symptoms of lowered mood, loss of appetite, poor sleep, lack of concentration and interest, lack of sense of enjoyment, occasional constipation and loss of libido.

Depressor anguli oris: muscle extending from the mandible (lower jaw) to the angle of the mouth.

Depressor labii inferioris: muscle extending from the mandible to the midline of the lower lip.

Dermal papilla: elevation at the base of the bulb which contains a rich blood supply to provide the hair with food and oxygen.

Dermatosis papulosa nigra (DPN): benign cutaneous condition that is common among black skins. It is characterised by multiple, small, hyperpigmented, asymptomatic papules.

Dermis: deeper layer of the skin found below the epidermis.

Desquamation: shedding of dead skin cells from the horny layer (stratum corneum).

Diabetes insipidus: rare metabolic disorder in which a person produces large quantities of dilute urine and is constantly thirsty.

Diabetes mellitus: carbohydrate metabolism disorder in which sugars are not oxidised to produce enough energy due to lack of the pancreatic hormone insulin.

Diaphragm: dome-shaped muscle of respiration that separates the thorax from the abdomen.

Diarrhoea: condition where there is frequent bowel evacuation or the passage of abnormally soft or liquid faeces.

Diastolic: static pressure against the arterial wall during rest or pause between heart contractions.

Diffusion: process in which small molecules move from areas of high concentration to those of lower concentration.

Digestion: process of breaking down food.

Distal convoluted tubule: tube with a series of twists leading to a straightened, collecting duct which leads to the pelvis of the kidney and on to the ureter.

Dorsiflexion: upward movement of the foot so that feet point upwards.

Duodenum: first of the three parts of the small intestine.

Dupuytren's contracture: forward curvature of the fingers (usually the ring and little fingers) caused by contracture of the fibrous tissue in the palm and fingers.

Dwarfism: hyposecretion of the growth hormone during childhood.

Dysmenorrhoea: painful and difficult menstruation.

Eccentric contraction: type of contraction when a muscle is stretched as it tries to resist a force pulling the bones of attachment away from one another.

Eccrine gland: simple, coiled, tubular sweat gland that opens directly on to the surface of the skin.

Ectopic pregnancy: development of a foetus at a site other than in the uterus. The most common type of ectopic pregnancy occurs in the fallopian tube.

Eczema: mild to chronic inflammatory skin condition characterised by itchiness, redness and the presence of small blisters that may be dry or weep if the surface is scratched. Eczema is not contagious but its cause may be genetic or due to internal and external influences.

Ejaculation: discharge of semen from the erect penis at the moment of sexual climax in the male.

Ejaculatory ducts: short tubes which join the seminal vesicles to the urethra.

Elastin: protein in the dermis which gives the skin its elasticity.

Emphysema: chronic, obstructive pulmonary disease in which the alveoli of the lungs become enlarged and damaged, reducing the surface area for the exchange of oxygen and carbon dioxide.

Emulsification: process by which fat globules are broken up into smaller droplets by the action of bile salts.

Endocardium: inner layer of the heart.

Endocrine gland: ductless gland that manufactures one or more hormones and secretes them directly into the bloodstream.

Endometriosis: condition in which tissue resembling the lining of the uterus (endometrium) is abnormally present in the pelvic cavity.

Endomysium: fine, connective tissue sheath that surrounds a single muscle fibre.

Endoplasmic reticulum: series of membranes continuous with the cell membrane. Cell's intracellular transport system, allowing movement of materials from one part of the cell to another.

Enzyme: chemical catalyst.

Ephelides: also known as freckles, these present as small, pigmented areas of skin and appear where there is excessive production of the pigment melanin (after exposure to sunlight).

Epidermis: outermost, superficial layer of the skin.

Epididymides: highly convoluted tubes that connect the testes to the vas deferens.

Epilepsy: neurological disorder which makes the individual susceptible to recurrent and temporary seizures.

Epimysium: fibrous, elastic tissue surrounding a muscle.

Erector pili muscle: small, smooth, weak muscle attached at an angle to the base of a hair follicle which makes the hair stand erect in response to cold.

Erector spinae: long, postural muscle in three bands either side of spine, attaching to the spine, ribcage and head.

Erythema: deadening of the skin due to the dilation of blood capillaries just below the epidermis.

Erythrocyte: red blood cells that transport the gases of respiration.

Ethmoid: skull bone forming part of the wall of the orbit, roof of the nasal cavity and part of the nasal septum.

Eversion: soles of the feet face outwards.

Expiration: act of breathing out air from the lungs.

Extension: straightening of a body part at a joint so that the angle between the bones is increased.

Extensor carpi radialis: muscle extending along the radial side of the posterior of the forearm.

Extensor carpi ulnaris: muscle extending along the ulnar side of the posterior of the forearm.

Extensor digitorum: muscle extending along the lateral side of the posterior of the forearm.

Extensor digitorum longus: long muscle of the lower leg that extends the toes.

Extensor hallicus longus: long muscle of the lower leg that extends the big toe.

Facial nerve: mixed nerve that conducts impulses to and from several areas in the face and neck. The sensory branches are associated with the taste receptors on the tongue and the motor fibres transmit impulses to the muscles of facial expression.

Faeces: solid or semi-solid mass of undigested food that is eliminated through the anus.

Fallopian tubes: tubes extending on the sides of the uterus, passing upwards and outwards to end near each ovary. Their function is to convey the ovum from the ovary to the uterus.

Fascia: fibrous, connective tissue that envelops muscles and organs.

Fasciculi: bundle of muscle fibres.

Fatty acids: end product of fat digestion.

Femur: long bone of the thigh.

Fertilisation: fusion of a spermatozoon and an ovum.

Fibrin: insoluble, fibrous protein formed from fibrinogen during blood coagulation.

Fibrinogen: plasma protein that is converted to fibrin during blood coagulation.

Fibroblast: cell found in reticular layer of dermis that forms new fibrous tissue.

Fibroid: abnormal growth of fibrous and muscular tissue, one or more of which may develop in the muscular wall of the uterus.

Fibromyalgia: chronic condition that produces musculoskeletal pain.

Fibrositis: inflammatory condition of the fibrous, connective tissues, especially in the muscle fascia (also known as muscular rheumatism).

Fibrous joint: immovable joint with tough, fibrous tissue between the bones.

Fibrous joint capsule: part of synovial joint reinforced by connective tissue to hold the bones together and enclose the joint.

Fibula: long bone situated on lateral side of tibia in lower leg.

Filtration: movement of water and dissolved substances across the cell membrane due to differences in pressure.

Fissure: crack in the epidermis, exposing the dermis.

Fixator: muscles that stabilise a bone to give a steady base from which the agonist works.

Flexion: bending of a body part at a joint so that the angle between the bones is decreased.

Flexor carpi digitorum: anterior muscle of the forearm extending from the medial end of the humerus, the anterior of the ulna and radius to the anterior surfaces of the second to fifth fingers.

Flexor carpi radialis: muscle of the forearm extending along the radial side of the anterior of forearm.

Flexor carpi ulnaris: anterior muscle of the forearm extending along the ulnar side of the anterior of the forearm.

Flexor digitorum longus: long muscle of the lower leg that flexes the toes.

Flexor hallicus longus: long muscle of the lower leg that flexes the big toe.

Follicle-stimulating hormone: in women, stimulates the development of the graafian follicle in the ovary which secretes the hormone oestrogen. In men, it stimulates the testes to produce sperm.

Folliculitis: bacterial infection which occurs in the hair follicles of the skin and appears as a small pustule at the base of a hair follicle. There is redness, swelling and pain around the hair follicle.

Fracture: breakage of a bone, either complete or incomplete.

Free edge: part of the nail plate that extends beyond the nail bed.

Frontal: skull bone forming the forehead.

Frontalis: muscle that extends over the front of the skull and the width of the forehead.

Frozen shoulder (adhesive capsulitis): chronic condition in which there is pain and stiffness and reduced mobility, or locking, of the shoulder joint.

Gall bladder: pear-shaped sac lying underneath the right lobe of the liver where bile is stored.

Gallstones: hard, pebblelike mass which is formed within the gall bladder.

Gastric juice: liquid secreted by the gastric glands of the stomach (main constituents are hydrochloric acid, mucus, rennin and pepsinogen).

Gastrin: hormone released by endocrine cells in the stomach wall and stimulated by the presence of food.

Gastrocnemius: large, superficial calf muscle with two bellies.

Genitalia: external reproductive organs.

Gigantism: abnormal growth causing excessive height, most commonly due to oversecretion of the growth hormone during childhood.

Gliding joint: type of synovial joint where two flat bones slide over one another.

Glomerulus: network of blood capillaries contained within the cuplike end (Bowman's capsule) of a nephron.

Glossopharyngeal nerve: mixed nerve that innervates structures in the mouth and throat.

Glucocorticoids: group of steroid hormones synthesised by the adrenal cortex, essential for the utilisation of carbohydrate, fat and protein by the body and for a normal response to stress.

Glucose: single sugar that serves as the primary source of cellular energy, the end product of carbohydrate digestion.

Gluteus maximus: large muscle covering the buttock.

Gluteus medius: medium-sized muscle of the buttocks.

Gluteus minimus: smallest of the buttock muscles.

Glycerol: clear, viscous liquid obtained by the breakdown of fats and the end product of fat digestion.

Glycogen: carbohydrate consisting of branched chains of glucose units, the principal form in which carbohydrate is stored in the body.

Golgi body: collection of flattened sacs within the cytoplasm near the nucleus and attached to the endoplasmic reticulum. It is the 'packaging and storage' department of the cell.

Gonadotrophic hormones: secreted from the anterior lobe of the pituitary gland, these hormones control the development and growth of the ovaries and testes.

Gracilis: long, straplike muscle that adducts the thigh.

Granular layer (stratum granulosum): layer of epidermis linking the living cells of the epidermis (basal and prickle-cell layers) to the dead cells above.

Greenstick fracture: partial fracture in which one side of the bone is broken and the other side bends (only occurs in children).

Growth hormone: secreted from the anterior lobe of the pituitary gland, controls the growth of long bones and muscles.

Gynaecomastia: enlargement of the breasts in the male due to either a hormone imbalance or hormone therapy.

Haemophilia: hereditary disorder in which the blood clots very slowly due to deficiency of either of two coagulation factors – Factor VIII (the antihaemophiliac factor) or Factor IX (the Christmas factor).

Haemorrhoids: condition with abnormal dilatation of veins in the rectum.

Hair: appendage of the skin which grows from a saclike depression in the epidermis called a hair follicle.

Hair bulb: enlarged part at the base of the hair root.

Hair root: part found below the surface of the skin.

Hair shaft: part of hair lying above the surface of the skin.

Hamstrings: group of three muscles situated on the posterior of the thigh.

Hay fever: allergic reaction involving the mucous passages of the upper respiratory tract and the conjunctiva of the eyes, caused by pollen or other allergens.

Headache: pain felt deep within the skull but excluding facial pain.

Heart: hollow organ made up of cardiac muscle tissue which lies in the thorax above the diaphragm and between the lungs. Its function is to maintain a constant circulation of blood throughout the body.

Heart attack (myocardial infarction): damage to the heart muscles which results from blockage of the coronary arteries.

Heartburn: burning sensation felt behind the sternum and often appearing to rise from the abdomen, up the oesophagus, towards or into the throat.

Hepatic arteries: blood vessels supplying blood to the liver.

Hepatic portal vein: blood vessels that drain blood from the liver.

Hepatitis: inflammation of the liver caused by viruses, toxic substances or immunological abnormalities.

Hepatitis A: is highly contagious and is transmitted by the faecal-oral route. It is transmitted by ingestion of contaminated food, water or milk.

Hepatitis B: also known as serum hepatitis. It lasts longer and is more serious than hepatitis A and can lead to cirrhosis, cancer of the liver and a carrier state.

Hepatitis C: can cause acute or chronic hepatitis. It can also lead to a carrier state and liver cancer. It is transmitted through blood transfusions or exposure to blood products.

Hernia: abnormal protrusion of an organ or part of an organ through the wall of the body cavity in which it normally lies.

Herpes simplex (cold sores): normally found on the face and around the lips. It begins as an itching sensation, followed by erythema and a group of small blisters which then weep and form crusts.

Herpes zoster (shingles): painful infection along the sensory nerves by the virus that causes chicken pox.

Hiatus hernia: most common type of hernia, occuring when part of the stomach is protruding into the chest.

High blood pressure: when the resting blood pressure is above normal, as consistently exceeding 160 mmHg systolic and 95 mmHg diastolic.

Hilus: medial border of the kidney where the renal blood vessels leave and enter the kidney.

Hinge joint: type of synovial joint where the rounded surface of one bone fits the hollow surface of another bone. Movement is only possible in one direction.

Hirsutism: presence of coarse, pigmented hair on the face, chest and upper back or abdomen in a female, due to excessive production of androgens.

Hodgkin's disease: malignant disease of the lymphatic tissues, usually characterised by painless enlargement of one or more groups of lymph nodes in the neck, armpit, groin, chest or abdomen.

Homeostasis: process by which the body maintains a stable internal environment for its cells and tissues.

Hormone: chemical messenger or regulator, secreted by an endocrine gland which reaches its destination by the bloodstream and has the power of influencing the activity of other organs.

Horny layer (stratum corneum): most superficial, outer layer of the skin, consisting of dead, flattened, keratinised cells.

Humerus: long bone of the upper arm.

Hydrochloric acid: strong acid present in a very dilute form in gastric juice.

Hyperhidrosis: excessive production of sweat affecting the hands, feet and underarms.

Hypertrophic disorders: refer to conditions which have resulted in an increase of size of a tissue or organ. This is caused by an enlargement of the cells.

Hypoglossal nerve: motor nerve that innervates the muscles of the tongue.

Hypoglycaemia: deficiency of glucose in the bloodstream, causing muscular weakness and incoordination, mental confusion and sweating. If severe it can lead to a hypoglycaemic coma.

Hypothalamus: small structure lying beneath the thalamus which governs many important homeostatic functions.

Hypothenar eminence: projection of soft tissue located on the ulnar side of the palm of the hand. It consists of three muscles – abductor digiti minimi manus, flexor digiti minimi manus and opponens digiti minimi.

Ileum: lowest part of the three portions of the small intestine.

Iliacus: large, fan-shaped muscle deeply situated in the pelvic girdle.

Ilium: largest and most superior pelvic bone.

Immunisation: artificial stimulation of the body into producing antibodies.

Immunity: body's ability to resist infection.

Impacted fracture: where one fragment of bone is driven into the other.

Impetigo: superficial, contagious, inflammatory disease caused by streptococcal and staphylococcal bacteria. It is commonly seen on the face and around the ears and features include weeping blisters which dry to form honey-coloured crusts.

Incontinence: condition in which the individual is unable to control urination voluntarily.

Inferior vena cava: main vein receiving blood from the lower parts of the body, below the diaphragm.

Infertility: inability in a woman to conceive or in a man to induce conception.

Infraspinatus: muscle that attaches to the middle two-thirds of the scapula, below the spine of the scapula at one end and the top of the humerus at the other.

Ingestion: act of taking food into the alimentary canal through the mouth.

Inguinal nodes: lymph nodes located in the groin. They drain lymph from the lower limbs, the external genitalia and the lower abdominal wall.

Inner root sheath: originates from the dermal papilla at the base of the follicle and grows upwards with the hair, shaping and contouring.

Innominate artery (brachiocephalic): short artery originating as the first large branch of the aortic arch.

Innominate vein (brachiocephalic): either of the two veins, one on each side of the neck, formed by the junction of the external jugular and subclavian veins.

Insertion: most movable part of a muscle.

Inspiration: act of breathing air into the lungs through the mouth and nose.

Insulin: hormone produced in the pancreas by the islets of Langerhans cells. It is important for regulating the amount of glucose in the blood.

Intercostal muscles: muscles that occupy the spaces between the ribs and are responsible for controlling some of the movements of the ribs.

Intestinal juice: released by the glands of the small intestine and completes the final breakdown of nutrients, including simple sugars to glucose and protein to amino acids.

Inversion: soles of the feet face inwards.

Irritable bowel syndrome: common condition in which there is recurrent abdominal pain with constipation and/or diarrhoea and bloating.

Ischium: bone forming the posterior part of the pelvic girdle.

Islets of Langerhans: small groups of cells scattered through the pancreas. They secrete the hormones insulin and glucagon.

Isometric contraction: when a muscle works without actual movements or changing length.

Isotonic contraction: when a muscles force is considered to be constant but the muscle length changes.

Jaundice: yellowing of the skin or whites of the eyes, caused by excessive bilirubin (bile pigment) in the blood.

Jejunum: middle part of the small intestine, connecting the duodenum to the ileum.

Joint: point where two or more bones or cartilage meet.

Jugular vein: major vein draining blood from the head and neck (divides into internal and external branches).

Keloid: overgrowth of an existing scar which grows much larger than the original wound.

Keratin: tough, fibrous protein found in the epidermis, hair and nails.

Keratinisation: process that cells undergo when they change from living cells with a nucleus to dead, horny cells without a nucleus.

Kidneys: pair of organs responsible for excretion of waste from blood; situated at the back of abdomen, one on each side of spine.

Kidney stones: deposits of substances found in the urine which form solid stones with the renal pelvis of the kidney, the ureter or the bladder.

Kyphosis: abnormally increased outward curvature of the thoracic spine.

Lacrimal: smallest of the facial bones, located close to the medial part of the orbital cavity.

Lacteal: intestinal lymphatic vessel.

Lactic acid: compound that forms in the cells as an end product of glucose metabolism in the absence of oxygen.

Lanugo hair: fine, soft hair found on a foetus.

Large intestine: part of the digestive system concerned with absorption of water from the material passed from the small intestine. It consists of the caecum, appendix, colon and rectum.

Larynx (voice box): short passage connecting the pharynx to the trachea.

Lateral longitudinal arch: arch of the foot which runs along the lateral side of the foot from the calcaneum bone to the end of the metatarsals.

Latissimus dorsi: wide muscle of the back, extending across the back of the thorax.

Lentigo: also known as liver spots, these are flat, dark patches of pigmentation which are found mainly in the elderly, on skin exposed to light.

Lesion: zone of tissue with impaired function as a result of damage by disease or wounding.

Leucocyte: white blood cell that aids the body's defence mechanism.

Leukaemia: cancer of blood-forming organs characterised by rapid growth and distorted development of leucocytes.

Levator anguli oris: muscle extending from the maxilla (upper jaw) to the angle of the mouth.

Levator ani: muscle forming the pelvic floor.

Levator labii superioris: muscle located towards the inner cheek, beside the nose, and extending from the upper jaw to the skin of the corners of the mouth and the upper lip.

Levator scapula: straplike muscle that runs almost vertically through the neck, connecting the cervical vertebrae to the scapula.

Ligament: dense, strong, flexible band of white, fibrous, connective tissue that links bones together at a joint.

Liver: largest gland in the body, occupying the top right portion of the abdominal cavity. It is concerned with regulation of blood sugar and amino acid levels, secretion of bile and detoxification of waste.

Loop of Henle: part of the kidney tubule that forms a loop, extending towards the centre of the kidney. It absorbs water and selected soluble substances back into the bloodstream.

Lordosis: abnormally increased inward curvature of the lumbar spine.

Low blood pressure: when the blood pressure is below normal and as a systolic blood pressure of 99 mmHg or less and a diastolic of less than 59 mmHg.

Lower limb: part of the skeleton containing the femur, patella, tibia, fibula, tarsals, metatarsals and phalanges.

Lumbar plexuses: spinal nerves located between the waist and the hip. They supply the front and sides of the abdominal wall and part of the thigh.

Lumbar vertebrae: five vertebrae of the lower back.

Lungs: cone-shaped, spongy organs situated in the thoracic cavity on either side of the heart. Their function is to facilitate the exchange of the gases oxygen and carbon dioxide.

Lunula: light-coloured, semicircular area of the nail, commonly called the half-moon, lying between the matrix and the nail plate.

Lupus erythematosus: chronic, inflammatory disease of connective tissue affecting the skin and various internal organs. It is an autoimmune disease and can be diagnosed by the presence of abnormal antibodies in the bloodstream.

Luteinizing hormone: in women it helps to prepare the uterus for the fertilised ovum. In men it acts on the testes to produce testosterone.

Lymph: transparent, colourless, watery liquid derived from tissue fluid.

Lymphatic capillaries: minute, blind-end tubes that commence in the tissue spaces of the body.

Lymphatic node: oval or bean-shaped structure that filters lymph.

Lymphatic vessels: tubes similar in structure to veins, with thin, collapsible walls and valves; responsible for transporting lymph through its circulatory pathway.

Lymphoma: malignant tumour of the lymph nodes.

Lysosome: round sacs present in the cytoplasm, containing powerful enzymes to destroy any part of the cell that is worn out.

Macule: small, flat patch of increased pigmentation or discolouration such as a freckle.

Malignant: term used to describe a tumour that invades and destroys the tissue in which it originates and can spread to other sites in the body via the bloodstream and lymphatic system.

Malignant melanoma: deeply pigmented mole which is life-threatening if it is not recognised and treated promptly. Its main characteristic is a blue-black nodule which increases in size, shape and colour and is most commonly found on the head, neck and trunk.

Mandible: only movable bone of the skull, forming the lower jaw.

Masseter: thick muscle in the cheek extending from the zygomatic arch to the outer corner of the mandible.

Mast cells: cells found in reticular layer of dermis which secrete histamine during an allergic reaction.

Mastication: process of chewing food.

Mastoid nodes: lymph nodes located behind the ear; they drain lymph from the skin of the ear and the temporal region of the scalp.

Matrix: area of mitotic activity of the hair cells located at the lower part of the hair bulb; area of nail situated immediately below the cuticle, the area where the living cells are produced; substance of a tissue or organ in which specialised structures are embedded.

Maxilla: largest bone of the face, forming the upper jaw.

Medial longitudinal arch: arch of the foot running along the medial side of the foot from the calcaneum bone to the end of the metatarsals.

Medulla: inner layer of hair; inner part of kidney, made up of conical-shaped sections called renal pyramids.

Meiosis: type of cell division that produces four daughter cells, each having half the number of chromosomes of the original cell.

Melanocyte-stimulating hormone (MSH): hormone synthesised and released by the pituitary gland. It stimulates the production of melanin in the basal cell layer of the skin.

Melanoma: cancerous growth of melanocytes.

Melatonin: hormone produced by the pineal gland. It is involved in the regulation of circadian rhythms, sleep/wake rhythms and is thought to influence the mood.

Meninges: special type of connective tissue with three layers that protects the brain and spinal cord.

Meningitis: inflammation of the meninges due to infection by viruses or bacteria.

Menopause: time when a women ceases to menstruate and is no longer able to bear children.

Menstrual cycle: periodic sequence of events in women in which an egg cell (ovum) is released from the ovary at four-weekly intervals until the menopause.

Mentalis: muscle radiating from the lower lip over the centre of the chin.

Metabolism: physiological processes to convert the food we eat and the air we breathe into the energy the body needs to function.

Metacarpals: five long bones in the palm of the hand.

Metastasis: spread of cancerous cells to other parts of the body.

Metatarsals: five long bones forming the dorsal surface of the foot.

Migraine: specific form of headache, usually unilateral (one side of the head), associated with nausea or vomiting and visual disturbances such as scintillating or zigzag light waves.

Milia: sebum trapped in a blind duct with no surface opening. They appear as pearly, white, hard nodules under the skin.

Mineral corticoid: hormone secreted by the adrenal cortex, acting on the kidney tubules, retaining salts in the body, excreting excess potassium and maintaining the water and electrolyte balance.

Mitochondria: oval-shaped organelles that lie within the cytoplasm. It is the site of the cell's energy production.

Mitosis: type of cell division when a single cell produces two genetically identical daughter cells.

Mixed/association neurone: type of nerve cell that links sensory and motor neurones, helping to form the complex pathways that enable the brain to interpret incoming sensory messages, decide on what should be done and send out instructions in response along motor pathways.

Mole: also known as a pigmented naevus. They appear as round, smooth lumps on the surface of the skin. They may be flat or raised and vary in size and colour, from pink to brown or black.

Motor nerve: type of nerve that carries impulses outwards from the central nervous system to bring about an activity in a muscle or gland.

Motor/efferent neurone: type of nerve cell that conducts impulses away from the brain and spinal cord to muscles and glands.

Motor neurone disease: progressive, degenerative disease of the motor neurones of the nervous system. It tends to occur in middle age and causes muscle weakness and wasting.

Motor point: point where the nerve supply enters the muscle.

Mucus: viscous fluid secreted by mucous membrane, acting as a protective barrier and lubricant.

Multiple sclerosis: disease of the central nervous system in which the myelin (fatty) sheath covering the nerve fibres is destroyed and various functions become impaired, including movement and sensations.

Muscle cramp: acute, painful contraction of a single muscle or group of muscles.

Muscle fatigue: loss of the ability of a muscle to contract efficiently due to insufficient oxygen, exhaustion of energy supply and the accumulation of lactic acid.

Muscle spasm: increase in muscle tension due to excessive motor nerve activity, resulting in a knot in the muscle.

Muscle tone: state of partial contraction of a muscle.

Muscular atrophy: wasting away of muscles due to poor nutrition, lack of use or a dysfunction of the motor nerve impulses.

Muscular dystrophy: progressively crippling disease in which the muscles gradually weaken and atrophy.

Myelin sheath: fatty, insulating sheath that covers the axon. Its function is to insulate the nerve and accelerate the conduction of nerve impulses along the length of the axon.

Myocardium: middle layer of heart.

Myofibrils: contractile filaments found within skeletal/voluntary muscle cells.

Myosin: most abundant protein found in muscle that plays an important role in muscle contraction.

Myositis: inflammation of a skeletal muscle.

Myxoedema: hypothyroidism in adult life. Symptoms include coarsening of the skin, intolerance to cold, weight gain and mental dullness.

Naevus: birthmark and clearly defined malformation of the skin; mass of dilated capillaries.

Nail bed: area of the nail below the nail plate that provides nourishment and protection for the nail.

Nail groove: deep ridges under the sides of the nail which guide it and help it to grow straight.

Nail plate: main visible part of the nail which rests on the nail bed and ends at the free edge.

Nail wall: fold of skin overlapping the sides of the nails to protect the edges of the nail plate from external damage.

Nasal: small facial bone forming bridge of nose.

Nasalis: muscle located at the sides of the nose.

Nasopharynx: upper part of the nasal cavity, behind the nose.

Neoplasm: any new or abnormal growth (any benign or malignant tumour).

Nephritis: general, non-specific term used to describe inflammation of the kidney.

Nephron: tiny blood-filtration unit inside kidney.

Neuralgia: attacks of pain along the entire course or branch of a peripheral sensory nerve. A common example is trigemenal neuralgia affecting the trigeminal nerve in the face.

Neuritis: inflammation/disease of a single or several nerves, caused by infection, injury or poison.

Neuroglia: also known as glial cells, a special type of connective tissue of the central nervous system, designed to support, nourish and protect the neurones.

Neurone: specialised nerve cell, designed to receive stimuli and conduct impulses.

Neurotransmitter: chemical substance released from nerve endings across synapses to other nerves.

Noradrenaline: hormone secreted by the medulla of the adrenal glands. It is closely related to adrenaline and helps the body create the conditions needed for rest through the parasympathetic nervous system.

Nose: organ which moistens, warms and filters the air and senses smell.

Nuclear membrane: perforated outer membrane enclosing the nucleus.

Nucleolus: dense, spherical structure inside nucleus, containing ribonucleic acid (RNA) to form ribosomes.

Nucleus: control centre of the cell. It regulates the cell's functions and contains genetic information (DNA).

Obliques (external and internal): muscles located at the sides of the waist (external obliques are superficial to the internal obliques).

Occipital: skull bone forming the back of the skull.

Occipital nodes: lymph nodes located at the base of the skull. They drain lymph from the back of scalp and the upper part of the neck.

Occipitalis: muscle found at the back of the head. It is attached to the occipital bone and the skin of the scalp.

Oculomotor nerve: mixed nerve that innervates both internal and external muscles of the eye and a muscle of the upper eyelid.

Oedema: abnormal swelling of body tissues due to an accumulation of tissue fluid.

Oestrogen: hormone controlling female sexual development.

Olfactory: pertaining to the sense of smell.

Olfactory nerve: sensory nerve of olfaction (smell).

Oncologist: physician who specialises in the study and practice of treating tumours.

Oncology: study and practice of treating tumours.

Optic nerve: sensory nerve of vision.

Orbicularis oculi: circular muscle that surrounds the eye.

Orbicularis oris: circular muscle that surrounds the mouth.

Organ: part of the body, composed of more than one tissue, that forms a structural unit responsible for a particular function.

Origin: most fixed point of a muscle.

Osmosis: process of the movement of water through the cell membrane from areas of low chemical concentration to areas of high chemical concentration.

Ossification: process of bone formation.

Osteoblast: bone-building cell.

Osteoclast: cartilage-destroying cell.

Osteocyte: mature bone cells.

Osteoporosis: brittle bones due to ageing and the lack of the hormone oestrogen which affects the ability to deposit calcium in the matrix of bone.

Outer root sheath: forms the follicle wall and provides a permanent source of growing cells (hair germ cells).

Ova: egg cells (plural of ovum).

Ovarian follicle: cavity in which an ovum is formed.

Ovaries: female sex glands lying on the lateral walls of the pelvis. They are concerned with the production of ova (eggs).

Ovum: egg cell.

Oxytocin: hormone secreted from the posterior lobe of the pituitary. It stimulates the uterus during labour and stimulates the breasts to produce milk.

Palatine: L-shaped bones which form the anterior part of the roof of the mouth.

Pancreas: gland extending from the loop of the duodenum to behind the stomach which secretes pancreatic juice.

Pancreatic amylase: enzyme that continues the breakdown of starch and has the same effects as salivary amylase.

Pancreatic juice: secreted by the pancreas into the duodenum and contains enzymes that continue the digestion of protein, carbohydrates and fats.

Pancreatic lipase: enzyme that breaks down lipids into fatty acids and glycerol.

Papillary layer: most superficial layer of the dermis, situated above reticular layer.

Papule: small, raised elevation on the skin, less than 1 cm in diameter, which may be red in colour. It often develops into a pustule.

Parasympathetic nervous system: one of the two divisions of the autonomic nervous system. It balances the action of the sympathetic division by working to conserve energy and create the conditions needed for rest and sleep.

Parietal: forms the upper sides of the skull and the back of the roof of the skull.

Parkinson's disease: damage to grey matter of brain known as basal ganglia. It causes involuntary tremors of limbs, with stiffness, rigidity and shuffling gait.

Parotid nodes: lymph nodes located at the angle of the jaw. They drain lymph from the nose, eyelids and ear.

Patella: kneecap.

Pectoralis major: thick, fan-shaped muscle covering the anterior surface of the upper chest.

Pectoralis minor: thin muscle that lies beneath the pectoralis major.

Pediculosis (lice): a contagious, parasitic infection where the lice live off the blood sucked from the skin. With head lice, nits may be found in the hair. They are pearl-grey or brown, oval structures found on the hair shaft close to the scalp. The scalp may appear red and raw due to scratching.

Pelvic girdle: bony structure to which bones of lower limb are attached (consists of ilium, ischium and pubis bones).

Pelvic nodes: lymph nodes located within the pelvic cavity, along the paths of the iliac blood vessels. They drain lymph from organs within the pelvic cavity.

Penis: male organ made of erectile tissue that carries the urethra, through which urine and semen are discharged.

Pepsin: enzyme in the stomach that begins the digestion of proteins.

Peptidases: group of digestive enzymes that split proteins in the stomach and intestine into their constituent amino acids.

Pericardium: outer layer of heart.

Perimysium: fibrous sheath that surrounds each bundle of muscle fibres.

Periosteum: thin membrane of connective tissue covering bone.

Peripheral nervous system (PNS): part of the nervous system containing all the nerves outside the central nervous system. It consists of cablelike nerves that link the central nervous system to the rest of the body.

Peristalsis: coordinated, rhythmical contractions of the circular and oblique muscles in the wall of the alimentary tract to break down food and move it along the alimentary canal.

Peroneus longus/brevis: muscles situated on the lateral aspect of the lower leg. They attach to the fibula and are plantar flexors of the foot.

Phalanges: finger and toe bones.

Pharynx: throat serving as an air and food passage.

Phlebitis: inflammation of the wall of a vein, most commonly seen in the legs as a complication of varicose veins.

Pineal gland: pea-sized mass of nerve tissue attached by a stalk in the central part of the brain. It secretes a hormone called melatonin and is involved in the regulation of circadian rhythms.

Piriformis: deeply seated, pelvic girdle muscle that is a deep lateral rotator of the thigh.

Pituitary: lobed structure attached by a stalk to the hypothalamus of the brain.

Pivot joint: type of synovial joint where a process of bone rotates in a socket.

Plantar flexion: downward movement of the foot so that feet face downwards towards the ground.

Platysma: superficial neck muscle that extends from the chest, up either side of the neck, to the chin.

Pleurisy: inflammation of the pleura of the lung. It presents as an intense, stabbing pain over the chest on breathing deeply.

Plexuses: network of nerves or blood vessels.

Pneumonia: inflammation of the lung caused by bacteria in which the alveoli become filled with inflammatory cells and the lung becomes solid. Symptoms include fever, malaise, headache, together with a cough and chest pain.

Polycystic ovary syndrome: condition caused by hyposecretion of the hormones oestrogen and progesterone in the female. It is characterised by cysts on the ovaries, cessation of periods, obesity, atrophy of the breasts, hirsutism and sterility.

Polypeptide: molecule consisting of three or more amino acids linked together by peptide bonds.

Popliteal nodes: lymph nodes located behind the knee. They drain lymph from the lower limbs through deep and superficial nodes.

Portal circulation: special branch of the systemic circulation which collects blood from the digestive organs and delivers it to the liver for processing, via the hepatic portal vein.

Port wine stain: also known as a deep capillary naevus. It is present at birth and may vary in colour from pale pink to deep purple. It has an irregular shape but is not raised above the skin's surface and is usually found on the face, but may also appear on other areas of the body.

Pregnancy: period in which a woman carries a developing foetus.

Premenstrual syndrome: term for the physical and psychological symptoms experienced from 3 to 14 days prior to the onset of menstruation.

Prickle-cell layer (stratum spinosum): binding and transitional layer between the stratum granulosum and the stratum germinativum.

Procerus: muscle located in between the eyebrows.

Progesterone: hormone responsible for preparing the uterus for pregnancy.

Prolactin: hormone secreted by the anterior lobe of the pituitary. It stimulates the secretion of milk from the breasts following birth.

Pronation: turning the hand so that the palm is facing downwards.

Pronator teres: anterior muscle of the forearm, crossing the anterior aspect of the elbow.

Proprioceptors: sensory nerve endings located in muscles and tendons that transmit information to coordinate muscular activity, organs and glands.

Prostate: male accessory sex gland situated just below the bladder. During ejaculation it secretes an alkaline fluid that forms part of semen.

Prostatitis: inflammation of the prostate gland which is usually caused by bacteria.

Proximal convoluted tubule: highly coiled, twisted tube that leads away from the Bowman's capsule of a nephron and into the loop of Henle.

Psoas: long, thick, deep pelvic muscle.

Psoriasis: chronic, inflammatory skin condition. Psoriasis may be recognised as the development of well-defined, red plaques, varying in size and shape and covered by white or silvery scales.

Pterygoids: facial muscles extending from the sphenoid bone in the skull to the mandible in the jaw.

Puberty: time at which the onset of sexual maturity occurs and the reproductive organs become functional.

Pubis: bone forming anterior of pelvic girdle.

Pulmonary arteries: two blood vessels (right and left) carrying deoxygenated blood from the right ventricle of the heart to the lungs.

Pulmonary circulation: circulatory system between the heart and lungs.

Pulmonary embolism: blood clot carried into the lungs.

Pulmonary veins: four blood vessels that carry oxygenated blood from the lungs and returns it to the left atrium of the heart.

Pulse: pressure wave that can be felt in the arteries which corresponds to the beating of the heart.

Pustule: small, raised elevation on the skin containing pus.

Quadratus lumborum: muscle located in the lower back/lumbar region.

Quadriceps: group of four muscles on the anterior aspect of the thigh.

Radius: long bone of the forearm (on thumb side of forearm).

Rectum: end part of large intestine where faeces are stored before defecation.

Rectus abdominis: long, flat muscle that extends bilaterally along the entire length of the front of the abdomen.

Reflex action/arc: rapid and automatic response to a stimulus without any conscious thought of the brain.

Renal artery: either of the two large arteries arising from the abdominal aorta and supplying the kidneys.

Renal pelvis: funnel-shaped cavity that collects urine from the renal pyramids in the medulla and drains it into the ureter.

Renal vein: blood vessels via which treated blood leaves the kidney.

Reticular fibres: fibres found in the reticular layer of dermis which help to maintain the skin's tone, strength and elasticity.

Reticular layer: deepest layer of the dermis, situated below the papillary layer.

Rhinitis: inflammation of the mucous membrane of the nose, causing a blocked, runny and stuffy nose.

Rhomboids: either of the two muscles situated in the upper part of the back between the thoracic vertebrae and the scapula.

Ribosome: tiny organelles made up of ribonucleic acid (RNA) and protein. Its function is to manufacture proteins for use within the cell.

Right lymphatic duct: short duct that lies in the root of the neck and collects lymph from the right.

Ringworm: fungal infection of the skin which begins as small, red papules that gradually increase in size to form a ring. Affected areas on the body vary in severity from mild scaling to inflamed, itchy areas.

Risorius: triangular-shaped muscle that lies horizontally on the cheek, joining at the corners of the mouth.

Rodent ulcer: malignant tumour which starts off as a slow-growing, pearly nodule, often at the site of a previous skin injury.

Rosacea: chronic inflammatory disease of the face in which the skin appears abnormally red. The condition is gradual, and begins with flushing of the cheeks and nose. As the condition progresses it can become pustular.

Rotation: movement of a bone around an axis (180 degrees).

Rupture: tearing of a muscle fascia or tendon.

Sacral plexuses: spinal nerves located at the base of the abdomen supply the skin and muscles and organs of the pelvis.

Sacral vertebrae (sacrum): five fused vertebrae forming a flat, triangular-shaped bone between the pelvic bones.

Saddle joint: type of synovial joint where the articulating surfaces of bone have both rounded and hollow surfaces so that the surface of one bone fits the complementary surface of the other.

Saliva: alkaline liquid secreted by the salivary glands and by mucous membrane of the mouth.

Salivary amylase: enzyme that commences the digestion of starch or carbohydrates in the mouth.

Sarcoma: general term for any cancer arising from muscle cells or connective tissues.

Sarcomere: basic contractile unit of which skeletal/voluntary muscle fibres are composed.

Sartorius: narrow, ribbon-like muscles which cross the anterior of the thigh.

Scabies: contagious, parasitic skin condition caused by the female mite which burrows into the horny layer of the skin where she lays her eggs.

Scapula: bone forming posterior of shoulder girdle.

Scar: mark left on the skin after a wound has healed.

Sciatica: lower back pain which can affect the buttock and thigh.

Scoliosis: lateral curvature of the vertebral column which may be to the left or right side.

Scrotum: paired sac that holds the testes and epididymides.

Sebaceous cyst: round, nodular lesion, with a smooth, shiny surface, which develops from a sebaceous gland.

Sebaceous gland: small, saclike pouches found all over the body, except for the soles of the feet and the palms of the hands. They produce an oily substance called sebum.

Seborrhoea: excessive secretion of sebum by the sebaceous glands. The glands are enlarged and the skin appears greasy, especially on the nose and the centre zone of the face. The condition may develop into acne vulgaris and is common at puberty, lasting for a few years.

Seborrhoeic dermatitis: mild to chronic inflammatory disease of hairy areas well supplied with sebaceous glands. Common sites are the scalp, face, axillae and in the groin.

Semen: fluid ejaculated from the penis at sexual climax.

Seminal vesicles: pair of male accessory sex glands that open into the vas deferens before it joins the urethra. They secrete most of the liquid component of semen.

Seminiferous tubules: long, convoluted tubules that make up the bulk of the testes.

Sensory/afferent neurone: type of nerve cell that receives stimuli from sensory organs and receptors and transmits the impulse to the spinal cord and brain.

Septum: partition dividing an anatomical structure such as the heart.

Serratus anterior: broad, curved muscle located on the side of the chest/ribcage, below the axilla.

Sex corticoids: hormones secreted by the adrenal cortex, controlling the development of the secondary sex characteristics and the function of the reproductive organs, including testosterone, oestrogen and progesterone.

Shin splints: soreness in the front of the lower leg due to straining of the flexor muscles used in walking.

Shoulder girdle: part of skeleton that connects the upper limbs with the thorax and consists of two scapulae and clavicles.

Simple fracture: clean break with little damage to surrounding tissues and no break in the overlying skin (also known as a closed fracture).

Sinoatrial (SA) node: built-in electrical system which sets the pace of the heart rate.

Sinusitis: condition involving inflammation of the paranasal sinuses. It is usually caused by a viral or bacterial infection or may be associated with a common cold or allergy.

Skeletal/voluntary muscle: striped in appearance, this type of muscle tissue is attached to the skeleton. It is responsible for the movement of bones.

Skin tag: small growths of fibrous tissue which stand up from the skin and sometimes are pigmented.

Small intestine: part of the digestive system where most of the processes of digestion and absorption of food take place. It consists of three parts – duodenum, jejunum and ileum.

Smooth/involuntary muscle: type of muscle tissue found in the walls of hollow organs such as the stomach, intestines, bladder, uterus and blood vessels.

Soleus: large, flat calf muscle, situated deep in gastrocnemius.

Somatic nervous system: division of the peripheral nervous system that consists of 31 pairs of spinal nerves and 12 pairs of cranial nerves.

Spasticity: increase in muscle tone and stiffness.

Specific immunity: type of immunity programmed genetically in the human body from birth. It includes mechanical barriers, chemicals, inflammation, phagocytosis and fever.

Sperm (spermatozoa): mature, male sex cells.

Sphenoid: skull bone located in front of the temporal bone.

Sphygmomanometer: instrument used to measure blood pressure.

Spider naevi: collection of dilated capillaries which radiate from a central papule.

Spina bifida: congenital defect of the vertebral column in which the halves of the neural arch of a vertebra fail to fuse in the midline.

Spinal cord: portion of the central nervous system enclosed in the vertebral column which extends from an opening at the base of the skull down to the second lumbar vertebra. Its function is to relay impulses to and from the brain.

Spinal nerves: set of 12 pairs of nerves originating from the spinal cord.

Spleen: largest of the lymphatic organs, located in left-hand side of the abdominal cavity between the diaphragm and the stomach. It is concerned with protection from disease and the manufacture of antibodies.

Splenius capitis: long, posterior neck muscle that extends from the spinous processes of C7–T3 to the mastoid process of the temporal bone and the occipital bone.

Splenius cervicus: long, posterior neck muscle that extends from the spinous processes of T3–T6 to the transverse processes of C1–C3.

Sprain: injury to a ligament caused by overstretching or tearing.

Squamous cell carcinoma: malignant tumour which arises from the prickle-cell layer of the epidermis.

Sternocleidomastoid: long muscle that lies obliquely across each side of the neck. Its fibres extend upwards from the sternum and clavicle at one end to the mastoid process of the temporal bone (at the back of the ear).

Stomach: muscular, J-shaped organ located between the oesophagus and the small intestine. It is concerned with the mechanical breakdown of food.

Strain: injury caused by excessive stretching or working of a muscle or tendon that results in a partial or complete tear.

Strawberry naevus: raised, red lump usually appearing on the face and growing rapidly in the first month of life. It usually disappears spontaneously before the child reaches the age of ten.

Stress: factor which affects physical or emotional health.

Stroke: blocking of blood flow to the brain by an embolus in a cerebral blood vessel.

Stye: acute inflammation of a gland at the base of an eyelash caused by bacterial infection. The gland becomes hard and tender and a pus-filled cyst develops at the centre.

Subclavian artery: either of the two arteries supplying blood to the neck and arms.

Subclavian veins: two main veins at the base of the neck into which the jugular veins empty blood.

Subcutaneous layer: thick layer of connective and adipose tissue found below the dermis.

Submandibular nodes: lymph nodes located beneath the jaw. They drain lymph from the chin, lips, nose, cheeks and tongue.

Subscapularis: muscle that attaches the inside surface of the scapula to the anterior of the top of the humerus.

Superficial cervical nodes: lymph nodes located at the side of the neck which drain lymph from the lower part of the ear and the cheek region.

Superior vena cava: main vein draining blood from the upper parts of the body (head, neck, thorax and arms).

Supination: turning the hand so that the palm is facing upwards.

Supinator: muscle that attaches to the lateral aspect of the lower humerus and the radius and supinates the forearm.

Supraspinatus: muscle is located in the depression above the spine of the scapula.

Suptratrochlear nodes: lymph nodes located in the elbow region. They drain lymph from the upper limbs and pass through to the axillary nodes.

Sympathetic nervous system: one of the two divisions of the autonomic nervous system. It prepares the body for expending energy and dealing with emergency situations.

Synapse: minute gap across which nerve impulses pass from one neurone to the next at the end of a nerve fibre.

Synergyst: muscles on the same side of a joint, that work together to perform the same movements.

Synovial cavity: space between the articulating bones of a synovial joint.

Synovial fluid: thick, lubricating fluid secreted by the synovial membrane.

Synovial joint: freely movable joint.

Synovial membrane: membrane that forms the sac enclosing a freely movable joint.

Synovitis: inflammation of a synovial membrane in a joint.

System: group of organs that work together to perform a specific function.

Systemic circulation: largest circulatory system that carries oxygenated blood from the left ventricle of the heart through to the aorta.

Systolic: pressure exerted on the arterial wall during active ventricular contraction.

Tarsals: seven irregular bones in the foot.

Telangiecstasia: term for dilated capillaries where there is persistent vasodilation of capillaries in the skin.

Telogen: short, resting stage of hair.

Temporal: skull bone forming the sides of the skull below the parietal bones and above and around the ears.

Temporalis: fan-shaped muscle situated on the side of the skull above and in front of the ear.

Temporomandibular joint tension (TMJ syndrome): collection of symptoms and signs produced by disorders of the temporomandibular joint (hinge joint between the mandible of the jaw and the temporal bone of the skull).

Tendinitis: inflammation of a tendon accompanied by pain and swelling.

Tendon: white, fibrous cords of connective tissue that attach muscles to a bone.

Tennis elbow: inflammation of the tendons that attach the extensor muscles of the forearm at the elbow joint.

Tensor fascia lata: leg muscle attached to the outer edge of the ilium of the pelvis which runs via the long fascia lata tendon to the lateral aspect of the top of the tibia.

Teres major: muscle that attaches to the bottom lateral edge of the scapula at one end and the back of the humerus (just below the shoulder joint) at the other.

Teres minor: muscle that attaches to the lateral edge of the scapula, above teres major at one end and into the top of the posterior of the humerus at the other.

Terminal hair: longer, coarser hairs, mostly pigmented hairs found on the scalp, under the arms, eyebrows, pubic regions, arms and legs.

Testes: pair of male sex organs that produce sperm.

Testosterone: principal male sex hormone.

Tetany: condition caused by a reduction in the blood calcium level. It presents as spasm and twitching of the muscles, particularly those of the face, hands and feet.

Thalamus: positioned on each side of the forebrain, the thalami are relay and interpretation stations for the sensory messages (except olfaction) that enter the brain before they are transmitted to the cortex.

Thenar eminence: eminence of soft tissue located on the radial side of the palm of the hand.

Thoracic cavity: collective group of body parts (sternum, ribs and thoracic vertebrae) which offer protection for the heart and lungs.

Thoracic duct: main collecting duct of the lymphatic system that collects lymph from the left side of the head and neck, left arm, lower limbs and abdomen. It drains into the left subclavian vein to return it to the bloodstream.

Thoracic nodes: lymph nodes located within the thoracic cavity and along the trachea and bronchi. They drain lymph from the organs of the thoracic cavity and from the internal wall of the thorax.

Thoracic vertebrae: 12 vertebrae of the mid-spine.

Thrombocyte: smallest cellular elements of blood cells, significant in the blood-clotting process, otherwise known as platelets.

Thrombosis: condition in which the blood changes from a liquid to a solid state and produces a blood clot.

Thymus: gland composed of lymphatic tissue located in the upper chest. It is important in the newborn baby in promoting the development and maturation of lymphocytes.

Thyroid: gland in the neck situated on either side of the trachea. The principal role is in regulating metabolism.

Thyroid-stimulating hormone (TSH): secreted from the anterior lobe of the pituitary gland, it controls the growth and activity of the thyroid gland.

Thyrotoxicosis: excessive amounts of thyroid hormones in the bloodstream, causing rapid heartbeat, sweating, tremors, anxiety, increased appetite, loss of weight and intolerance to heat.

Thyroxine (T4): one of the hormones synthesised and secreted by the thyroid gland.

Tibia: long bone situated on anterior and medial side of lower leg.

Tibialis anterior: muscle on the anterior and lateral side of the lower leg, extending from the tibia to the metatarsals.

Tibialis posterior: muscle on the posterior of the lower leg, extending from the tibia to the metatarsals.

Tinea capitis: type of ringworm and a fungal infection of the scalp. It appears as painless, round, hairless patches on the scalp. Itching may be present and the lesion may appear red and scaly.

Tinea pedis (athlete's foot): highly contagious, fungal condition which is easily transmitted in damp, moist conditions such as swimming pools, saunas and showers. It appears as flaking skin between the toes which becomes soft and soggy.

Tissue: group of similar cells that perform a certain function.

Tissue (interstitial) fluid: intercellular fluid located between cells other than blood cells.

Tonsils: lymphatic tissue located in the oral cavity and the pharynx.

Torticollis: condition in which the neck muscles (sternomastoids) contract involuntarily. It is commonly called wryneck.

Trachea: windpipe that passes down into the thorax and connects the larynx with the bronchi.

Transverse arch: arch of the foot which runs between the medial and lateral aspect of the foot and is formed by the navicular, three cuneiforms and the bases of the five metatarsals.

Transversus abdominis: deep abdominal muscle that runs transversely across the abdomen.

Trapezius: large, triangular-shaped muscle in the upper back that extends horizontally from the base of the skull (occipital bone) and the cervical and thoracic vertebrae to the scapula.

Triangularis: triangular-shaped muscle located below the corners of the mouth.

Triceps: muscle on the posterior of the upper arm.

Trigeminal nerve: mixed nerve (containing motor and sensory nerves).

Trimester: any one of the three successive three-month periods (first, second and third) into which a pregnancy may be divided.

Triodothyronine (T3): one of the hormones synthesised and secreted by the thyroid gland.

Trochlear nerve: smallest of the cranial nerves. It is a motor nerve that innervates the superior oblique muscle of the eyeball which helps you look upwards.

Trypsin: enzyme that continues the digestion of proteins by breaking down peptones into smaller peptide chains and is secreted by the pancreas.

Tuberculosis (TB): infectious disease caused by the bacillus (bacteria) Mycobacterium tuberculosis. Main transportation is via droplet infection and most common site for the bacilli to spread is in the lungs.

Tumour: abnormal growth of tissue which may be benign or malignant. Tumours may be cancerous and sometimes fatal or they may be quite harmless.

Turbinate: layers of bone located either side of the outer walls of the nasal cavities.

Ulcer: break or open sore in the skin extending to all its layers.

Ulcer (peptic, duodenal or gastric): break in the lining of the alimentary tract which fails to heal and is accompanied by inflammation. This condition can present with increased acidity, epigastric pain and heartburn.

Ulna: long bone of the forearm on little finger side of forearm.

Upper limb: part of the skeleton containing the humerus, radius, ulna, carpals, metacarpals and phalanges.

Ureter: either of a pair of muscular tubes that convey urine from the pelvis of kidney to the bladder.

Urethra: tube that conducts urine from the bladder to the exterior. In the male it also serves as a conducting channel for semen.

Urinary bladder: pear-shaped sac which lies in the pelvic cavity where urine is collected and stored temporarily.

Urinary tract infection: bacterial infection of one or more of the structures of the urinary system.

Urine: fluid excreted by the kidneys, containing water, urea, uric acid and creatinine.

Urticaria: also known as 'hives'. It is an itchy rash resulting from the release of histamine by mast cells.

Uterus: small, hollow, pear-shaped organ situated behind the bladder and in front of the rectum. It is the area in which an embryo grows and develops into a foetus.

Vacuole: empty space within the cytoplasm containing waste materials or secretions formed by the cytoplasm.

Vagina: lower part of female reproductive tract. It is a muscular tube lined with mucous membrane, connecting the cervix of uterus to the exterior.

Vagus nerve: unlike the other cranial nerves, it has branches to numerous organs in the thorax and abdomen as well as the neck.

Varicose veins: veins which have become distended, lengthened and swollen.

Vas deferentia: pair of ducts that conduct sperm from the epididymis to the urethra on ejaculation.

Vein: type of blood vessels with thinner, muscular, elastic walls and valves. They carry blood towards the heart.

Vellus hair: soft, downy hair found all over the face and body, except for the palms of the hands, soles of the feet, eyelids and lips.

Ventricle: lower chamber of heart.

Vertebral arteries: main division of the subclavian artery, arising from the subclavian arteries in the base of the neck and passing upwards where they unite to form a single basilar artery.

Vertebral veins: main veins descending from the transverse openings (or foramina) of the cervical vertebrae and entering deep structures of the neck such as the vertebrae and muscles.

Vesicle: small, saclike blister.

Vestibulocochlear nerve: sensory nerve that transmits impulses generated by auditory stimuli and stimuli related to balance and movement.

Villi: short, fingerlike processes that project from membranous surfaces such as from the wall of the small intestine.

Virilism: masculinisation in the female from the development of a combination of increased body hair, muscle bulk, deepening of the voice and male psychological characteristics.

Vitiligo: areas of the skin lacking pigmentation due to the basal cell layer of the epidermis no longer producing melanin. The cause of vitiligo is unknown.

Vomer: single bone at the back of the nasal septum.

Vulva: female external genitalia.

Wart: benign growth on the skin caused by infection with the human papilloma virus.

Weal: raised area of skin containing fluid which is white in the centre, with a red edge.

Whiplash: condition produced by damage to the muscles, ligaments, intervertebral discs or nerve tissues of the cervical region by sudden hyperextension and/or flexion of the neck.

Zygomatic: facial bone forming the cheeks.

Zygomatic major and minor/zygomaticus: muscles lying in the cheek area, extending from the zygomatic bone to the angle of the mouth.

index

Note: the reader is recommended to consult the glossary, where one can find definitions for very many of the terms listed in this index.